Introduction to Music Therapy Pr

Rachael Finnerty MTA MMT MA
Fall 2018 Wednesdays 7-10pm LRW B1007
Office Hours by Appointment. Pre-requisite: Registration in Level II or above

This course offers an introduction to the literature and practice of music therapy with an emphasis on the diversity of music therapy applications such as bio-medical, psychoanalytical, behavioural and rehabilitation.

Attendance is required to fully understand the material

Quiz 1 Classes 1&2 Opens 9am Fri.Sept.14 & Closes 5pm Mon. Sept 17	Quiz 2 Classes 3&4 Opens 9am Fri.Sept.28 & Closes 5pm Mon. Oct 1	Quiz 3 Classes 5&6 Opens 9am Fri.Oct.19 & Closes 5pm Mon. Oct. 22	MidTerm Classes 1-6 Will be held during class time Oct 24	Quiz 4 Classes 7&8 Opens 9am Fri.Nov.9 & Closes 5pm Mon. Nov. 12	Quiz 5 Classes 9&10 Opens 9am Fri.Nov 23 & Closes 5pm Mon. Nov.26	Class 13 Review for Exam Final Exam During Exam Period
6%	6%	6%	30%	6%	6%	40%

Over 10% of the final grade will be accounted for by November 9

Quizzes will be on-line on Avenue – If a quiz is missed, the next quiz will be weighted an additional 6%. If Quiz 4 is missed, the final exam will be weighted an additional 6%.
The second quiz missed will be scored 0.
Four days are provided to complete the quiz – Quiz will not be made available after the allotted time. **No Exceptions**

Class 1/Sept 5th   **Overview of Music Therapy** Music Therapy in Canada, Music Therapy Interventions Courseware p.155-160 (Readings on Avenue)
Class 2/Sept 12th   **History of Music Therapy** Courseware p.1-20
Class 3/Sept 19th **Music Therapy Approaches & Models** Courseware p.21-46
Class 4/Sept 26th **Songwriting & Improvising as Music Therapy interventions** Courseware p.47-56
Class 5/Oct 3rd **Music Therapy and Acquired Brain Injury #musicitsscience** Class Slides

READING WEEK No Class Wed Oct 10th	
Class 6/ Oct 17 **Music Therapy and the Elderly & use of pre-composed music as a Music Therapy intervention** Courseware p.57-68	
Class 7/Oct 24 **Mid Term (Classes 1-6)**	
Class 8/ Oct 31 **Music Therapy and Autism Spectrum Disorder** Courseware p.77-96	
Class 9/ Nov 7th **Music Therapy and Mental Health Music Therapy & Anxiety** Courseware p.97-128 Readings on Avenue	
Class 10/ Nov 14th **Community Music Therapy & Music Therapy for "the well"** Readings on Avenue	
Class 11/ Nov 21st **Music Therapy & Palliative Care** Courseware p.129-154	
Class 12/ Nov 28 **Music Therapy in NICU & Music Therapy Research** Courseware p.161-176 & Reading on Avenue	
Class 13 Review for Final Exam	

Courseware

Introduction to Music Therapy 2MT3 Fall 2018

Policy for missed work
First quiz missed: Next Quiz 12%
Second quiz missed will be marked 0
If quiz 4 is missed, final exam will be weighted an additional 6%.
4 days are allotted to complete each quiz – **no extensions** will be provided.

Policies and Statements:

Introduction to Music Therapy Practice - Music 2MT3 (Co2)
Rachael Finnerty MTA MMT MA finnerr@mcmaster.ca
Fall 2018 Tuesdays 7-10pm JHE 264
Office Hours by Appointment. Pre-requisite: Registration in Level II or above

This course offers an introduction to the literature and practice of music therapy with an emphasis on the diversity of music therapy applications such as bio-medical, psychoanalytical, behavioural and rehabilitation.

Attendance is required to fully understand the material

Quiz 1 Classes 1&2 Opens 9am Fri.Sept.14 & Closes 5pm Mon. Sept 17	Quiz 2 Classes 3&4 Opens 9am Fri.Sept.28 & Closes 5pm Mon. Oct 1	Quiz 3 Classes 5&6 Opens 9am Fri.Oct.19 & Closes 5pm Mon. Oct. 22	MidTerm Classes 1-6 Will be held during class time Oct 23	Quiz 4 Classes 7&8 Opens 9am Fri.Nov.9 & Closes 5pm Mon. Nov. 12	Quiz 5 Classes 9&10 Opens 9am Fri.Nov 23 & Closes 5pm Mon. Nov.26	Class 13 Review for Exam Final Exam During Exam Period
6%	6%	6%	30%	6%	6%	40%

Over 10% of the final grade will be accounted for by November 9

Quizzes will be on-line on Avenue – If a quiz is missed, the next quiz will be weighted an additional 6%. If Quiz 4 is missed, the final exam will be weighted an additional 6%.
The second quiz missed will be scored 0.
Four days are provided to complete the quiz – Quiz will not be made available after the allotted time. **No exceptions**

Class 1/Sept 4th   **Overview of Music Therapy** Music Therapy in Canada, Music Therapy Interventions Courseware p.155-160 (Readings on Avenue)
Class 2/Sept 11th   **History of Music Therapy** Courseware p.1-20
Class 3/Sept 18th **Music Therapy Approaches & Models** Courseware p.21-46
Class 4/Sept 25th **Songwriting & Improvising as Music Therapy interventions** Courseware p.47-56
Class 5/Oct 2rd **Music Therapy (NMT) and Acquired Brain Injury #musicitsscience** Class Slides

READING WEEK No Class Tues Oct 9th	
Class 6/ Oct 17	
Music Therapy and the Elderly & use of pre-composed music as a Music Therapy intervention Courseware p.57-68	
Class 7/Oct 23 **Mid Term (Classes 1-6)**	
Class 8/ Oct 30	
Music Therapy and Autism Spectrum Disorder Courseware p.77-96	
Class 9/ Nov 6th	
Music Therapy and Mental Health **Music Therapy & Anxiety** Courseware p.97-128 Readings on Avenue	
Class 10/ Nov 13th	
Community Music Therapy & Music Therapy for "the well" Readings on Avenue	
Class 11/ Nov 20st	
Music Therapy & Palliative Care Courseware p.129-154	
Class 12/ Nov 27	
Music Therapy in NICU & Music Therapy Research Courseware p.161-176 & Reading on Avenue	
Class 13 / Dec 4th Review for Final Exam	

Courseware

Introduction to Music Therapy 2MT3 Fall 2018

Policy for missed work
First quiz missed: Next Quiz 12%
Second quiz missed will be marked 0
If quiz 4 is missed, final exam will be weighted an additional 6%.
4 days are allotted to complete each quiz – **no extensions** will be provided.

Policies and Statements:

MUSIC 2MT3

TABLE OF CONTENTS & ACKNOWLEDGEMENTS

PAGE

Music Therapy: Historical Perspective — 1
    Davis, W. B. and Gfeller, K. E
    <u>An Introduction to Music Therapy: Theory and Practice</u>,
    Davis, W.B. et al.
    © 2008 American Music Therapy Association
    Reprinted with permission.

Analytically Oriented Music Therapy - The Priestley Model — 21
    Wigram, T.
    <u>A Comprehensive Guide to Music Therapy: Theory, Clinical Practice, Research and Training</u>, Wigram, T. et al.
    © 2002 Jessica Kingsley Publishers
    Reprinted with permission.

Basic Therapeutic Methods and Skills — 31
    Wigram, T.
    <u>Improvisation: Methods and Techniques for Music Therapy Clinicians, Educators and Students</u>, Wigram, T.
    © 2004 Jessica Kingsley Publishers
    Reprinted with permission.

Songwriting to Explore Identity Change and Sense of Self-concept Following Traumatic Brain Injury — 47
    Baker, F. et al.
    <u>Songwriting: Methods, Techniques and Clinical Applications for Music Therapy Clinicians, Educators and Students</u>, Baker, F. & Wigram, T., eds.
    © 2005 Jessica Kingsley Publishers
    Reprinted with permission.

Music Therapy with the Elderly — 57
    Aldridge, D.
    <u>Music Therapy Research and Practice in Medicine: From Out of the Silence</u>, Aldridge, D.
    © 1996 Jessica Kingsley Publishers
    Reprinted with permission.

Preverbal Communication Through Music to Overcome a Child's Language Disorder   69
    Oldfield, A.
    <u>Case Studies in Music Therapy</u>, Bruscia, K.E. ed.
    © 2006 Barcelona Publishers
    Reprinted with permission.

Individuals with Autism and Autism Spectrum Disorder (ASD)   77
    Adamek M. S. et al.
    <u>An Introduction to Music Therapy: Theory and Practice</u>,
    Davis, W.B. et al.
    © 2008 American Music Therapy Association
    Reprinted with permission.

Integrated Music Therapy with a Schizophrenic Woman   97
    Perilli, G.G.
    <u>Case Studies in Music Therapy</u>, Bruscia, K.E. ed.
    © 2006 Barcelona Publishers
    Reprinted with permission.

Performance in Music Therapy: Experiences in Five Dimensions   111
    Jampel, P. F.
    <u>Voices: A World Forum for Music Therapy</u>, 11.1
    © 2011 VOICES:A World Forum for Music Therapy
    Reprinted with permission.

Music Therapy in Hospice and Palliative Care   129
    Walker, J. and Adamek, M.
    <u>An Introduction to Music Therapy: Theory and Practice</u>,
    Davis, W.B. et al.
    © 2008 American Music Therapy Association
    Reprinted with permission.

The Difference Between Music and Music Therapy   155
    Finnerty, R.P.
    <u>Music Therapy As An Intervention for Pain Perception</u>,
    Finnerty, R.P.
    © 2006 R.P. Finnerty
    Reprinted with permission.

CHAPTER 2

MUSIC THERAPY: HISTORICAL PERSPECTIVE

William B. Davis
Kate E. Gfeller

CHAPTER OUTLINE

MUSIC THERAPY IN PRELITERATE CULTURES
MUSIC AND HEALING IN EARLY CIVILIZATIONS
USES OF MUSIC IN ANTIQUITY: HEALING RITUALS
MUSIC AND HEALING IN THE MIDDLE AGES AND RENAISSANCE
MUSIC THERAPY IN THE UNITED STATES
    18th-Century Writings on Music Therapy
    Literature from the 19th Century
    Music Therapy in 19th-Century Educational Institutions
    Early 20th-Century Music Therapy
THE DEVELOPMENT OF THE MUSIC THERAPY PROFESSION

Scholars from diverse disciplines, including anthropology, psychology, musicology, and physiology, have long questioned why music has been in our behavioral repertoire for thousands of years (Hodges, 1996; Winner, 1982). Music has no apparent survival value, yet it has been an important part of all cultures, past and present. Music has been called "the universal language" and "the greatest good that mortals know." Throughout recorded time, it has been credited with the power to "solace the sick and weary" and to express unspoken emotions (Stevenson, 1967). It is remarkable that music has claimed such a valued role throughout history. This chapter will discuss the role of music in preliterate cultures, the relationship between music and healing during the advent of civilization, the early practice of music therapy in the United States, and the development of the music therapy profession.

MUSIC THERAPY IN PRELITERATE CULTURES

Preliterate societies are those which possess no system of written communication. Early nomadic people banded together in small groups for survival and eked out a living as hunters and food gatherers. They had no agriculture, political structure, or permanent housing. These small groups developed distinct customs and rituals that set them apart from other similar groups. We can only speculate about the musical component of prehistoric life, but we can gain some clues by studying how music is used in preliterate cultures that exist today. This knowledge helps us to understand human response to music and provides some background about the close relationship between music and healing (Nettl, 1956).

Members of preliterate cultures generally believe they are controlled by magical forces and surrounded by an evil, unpredictable environment. To remain healthy, they feel compelled to obey a complex set of regulations that protect them from the hostile forces of nature and their fellow human beings. They perceive magic as an integral part of a healthy and peaceful life (Sigerist, 1970).

Members of preliterate cultures believe in the power of music to affect mental and physical well-being. Music is often connected with supernatural forces. For example, among certain preliterate societies, the songs used in important rituals are thought to have come from superhuman or unearthly sources (Merriam, 1964; Sachs, 1965). These songs, with their unexplainable powers, are used for entreating the gods and in all activities requiring extraordinary assistance, such as in religious or healing rites.

In some preliterate societies, an ill person is viewed as a victim of an enemy's spell. He or she is blameless and thus enjoys special treatment from the group. In other societies, however, it is believed that a person suffers illness to atone for sins committed against a tribal god. As long as the afflicted member continues to contribute to the well-being of the family and community, status does not change. If the person becomes too ill to uphold social responsibilities, he or she is considered an outcast and ostracized. In these cultures, the cause and treatment of disease is primarily determined by the "medicine man," who often applies elements of magic and religion in order to exorcise the malevolent spirit or demon from the patient's body. The type of music used is determined by the nature of the spirit invading the body.

Because of slightly different concepts of disease among preliterate societies, the role of the musician/healer and style of music vary. In most instances, the tribal musician/healer holds a place of importance within society. It is this person's duty not only to determine the cause of the disease, but also to apply the appropriate treatment to drive the spirit or demon from the patient's body. Sometimes, music functions as a prelude to the actual healing ceremony. Drums, rattles, chants, and songs may be used during the preliminary ritual and also throughout the actual ceremony (Sigerist, 1970). It is important to note that the musician/healer usually does not act alone. Preliterate societies recognize the power of the group and include family and society members in the ritual. Healing séances or choruses provide spiritual and emotional support in order to facilitate a quick recovery (Boxberger, 1962). As noted earlier, we must speculate about the customs of preliterate societies from ancient times. If, as many scholars believe, that current practices among contemporary preliterate societies offer a "window" into the past, then is likely that music was an important part of healing ceremonies very early in human history.

MUSIC AND HEALING IN EARLY CIVILIZATIONS

The hunters and food collectors of preliterate cultures predominated for about 500,000 years. The advent of agriculture 8,000–10,000 years ago led to a more stable existence, the growth of larger populations, and the rise of civilization. Civilization is characterized by the evolution of written communication, the growth of cities, and technological achievement in areas that include science and medicine. It is a way of life for a large group of people living in a more or less permanent alliance with a particular set of customs and view of nature. The first civilizations appeared between 5000 and 6000 B.C. in an area that is now Iraq and became firmly established by 3500 B.C. Music played an important part in "rational medicine" during this time as well as in magical and religious healing ceremonies.

USES OF MUSIC IN ANTIQUITY: HEALING RITUALS

With the advent of civilization, the magical, religious, and rational components of medicine began to develop along separate lines. In ancient Egypt (c. 5000 B.C.), these elements existed side by side, but healers generally based a treatment philosophy on only one. Egyptian music healers enjoyed a privileged existence, due to their close relationship with priests and other important government leaders. Egyptian priest-physicians referred to music as medicine for the soul and often included chant therapies as part of medical practice (Feder & Feder, 1981).

During the height of the Babylonian culture (c. 1850 B.C.), disease was viewed within a religious framework. The sick person suffered as penance for sins committed against a god and was viewed by society as an outcast. Treatment, if offered, consisted of religious ceremonies to placate the offended deity (Sigerist, 1970). Healing rites often included music.

Music was regarded as a special force over thought, emotion, and physical health in ancient Greece. In 600 B.C., Thales was credited with curing a plague in Sparta through musical powers (Merriam, 1964). Healing shrines and temples included hymn specialists, and music was prescribed for emotionally disturbed individuals (Feder & Feder, 1981). The use of music for curing mental disorders reflected the belief that it could directly influence emotion and develop character. Among the notables of Greece who subscribed to the power of music were Aristotle, who valued it as an emotional catharsis; Plato, who described music as the medicine of the soul; and Caelius Aurelianus, who warned against indiscriminate use of music against madness (Feder & Feder, 1981).

By the 6th century B.C., rational medicine had almost completely replaced magical and religious rites in Greece. Although a minority still attributed illness to supernatural powers, the majority supported rational investigation into the causes of disease. For the first time in history, the study of health and disease was based on empirical evidence (Sigerist, 1970).

The predominant explanation of health and disease became the theory of the four cardinal humors. This theory was described by Polybus, son-in-law of Hippocrates, in his treatise, "On the Nature of Man," circa 380 B.C. The four humors were blood, phlegm, yellow bile, and black bile, and each element contained a unique quality. Good health was the result of maintaining a balance among the four humors, whereas an imbalance of two or more elements led to illness. Sick individuals were considered to be inferior. With only slight modification, this theory influenced medicine for the next 2,000 years, becoming most important during the Middle Ages.

MUSIC AND HEALING IN THE MIDDLE AGES AND RENAISSANCE

Although much of the splendor of classical Greece was lost during the Middle Ages, this time period (c. 476–1450 A.D.) represents an important connection between antiquity and the present. After the fall of the Roman Empire, Christianity became a major force in Western civilization. The influence of Christianity prompted a change in attitudes toward disease. Contrary to earlier thinking, a sick person was neither inferior nor being punished by gods. As Christianity spread throughout Europe, societies began to care for and treat their sick members. Hospitals were established to provide humanitarian care to people with physical ailments. Sufferers of mental illness, however, were not as fortunate. Mentally ill people were believed to be possessed by demons and were often incarcerated and abused (Boxberger, 1962).

Although Christian beliefs heavily influenced attitudes toward disease during the Middle Ages, the practice of medicine was still based on the theory of the four humors developed during Greek civilization. This framework also provided the basis for the role of music in treating illness. Numerous statesmen and philosophers believed in the curative powers of music, including Boethius, who claimed that music either improved or degraded human morals. Cassiodorus, like Aristotle, viewed music as a potent type of catharsis, whereas St. Basil advocated it as a positive vehicle for sacred emotion. Many believed hymns to be effective against certain unspecified respiratory diseases (Strunk, 1965).

During the Renaissance, advances in anatomy, physiology, and clinical medicine marked the beginning of the scientific approach to medicine. Despite developments in the laboratory, however, treatment of disease was still based on the teachings of Hippocrates and Galen and a sophisticated interpretation of the four humors. During this period, there was some integration of music, medicine, and art. For example, it was not unusual to find writings, such as those of Zarlino (a musician) and Vesalius (a physician), that touched on the relationship between music and medicine (Boxberger, 1962).

Music during the Renaissance was not only used as a remedy for melancholy, despair, and madness, but also prescribed by physicians as preventive medicine. Properly dispensed music was recognized then, as it is today, as a powerful tool to enhance emotional health. For those who could afford the luxury of attending live performances, music helped to maintain a positive outlook on life. Optimism was particularly important during this time, because Europe was being ravaged by epidemics that sometimes decimated entire villages (Boxberger, 1962).

During the Baroque period (1580–1750), music continued to be linked with the medical practice of the day, based as before on the theory of the four humors. In addition, the theory of temperaments and affections by Kircher (1602–1680) provided a fresh viewpoint on the use of music in the treatment of disease. Kircher believed that personality characteristics were coupled with a certain style of music. For example, depressed individuals responded to melancholy music. Cheerful people were most affected by dance music, because it stimulated the blood (Carapetyan, 1948). Thus, it became necessary for the healer to choose the correct style of music for treatment. Supporting the use of music to treat depression, Burton, in his *Anatomy of Melancholy*, stated that "besides that excellent power it hath to expel many other diseases, it is a sovereign remedy against Despair and Melancholy, and will drive away the Devil himself" (Burton, 1651). Other writers, such as Shakespeare and Armstrong, also included numerous examples of music as therapy in their plays and poems (Davis, 1985).

By the late 18th century, music was still advocated by European physicians in the treatment of disease, but a definite change in philosophy was underway. With increased emphasis on scientific medicine, physicians and scientists began to investigate scientifically supported explanations for illness, rather than religious or superstitious explanations. Consequently, the use of music to "soothe" the gods would no longer be consistent with contemporary views on illness and healing. While many aspects of medicine from that time period would be viewed through contemporary eyes as naive or simply wrong, there was important progress regarding scientifically based medical treatment. During this time period, music was relegated to special cases and applied by only a few physicians who viewed treatment from a holistic (multitherapeutic) framework. The belief that music influenced mood was one way in which music was considered relevant to medical practice at this time. This shift in medical practice and the role of music within healing practices were evident during the growth and development of music therapy in the United States.

MUSIC THERAPY IN THE UNITED STATES

The practice of music therapy in the United States has a long, storied history. Although music therapy as a profession became organized only in the 20th century, music has been used in this country to treat physical and mental ailments since the late 18th century.

18th-Century Writings on Music Therapy

The earliest known reference to music therapy in the United States was an unsigned article in *Columbian Magazine* in 1789. The article, entitled "Music Physically Considered," presented basic principles of music therapy that are still in use today and provided evidence of music therapy practice in Europe. Mainly using the ideas of Descartes (a French philosopher), the anonymous author developed a case for using music to influence and regulate emotional conditions. An interesting conclusion drawn by the author was that a person's mental state may affect physical health. The author also asserted that music, because of its effect on emotions, was a proven therapeutic agent. One other important point in this article was the author's advice that the skilled use of music in the treatment of disease required a properly trained practitioner. This advice is as pertinent now as it was in 1789 (Heller, 1987).

Another article published during this period was entitled "Remarkable Cure of a Fever by Music: An Attested Fact." This article was published in 1796 in the *New York Weekly Magazine*. The anonymous author described the case of an unnamed French music teacher who suffered from a severe fever. After nearly two weeks of constant distress, a concert was performed at the patient's request. His symptoms reportedly disappeared during the performance but returned upon its conclusion. The music was repeated throughout the man's waking hours, resulting in the suspension of his illness. In two weeks' time, the music teacher recovered completely.

Both authors based their conclusions of the music's effectiveness on anecdotal rather than scientific evidence. Such claims lack credibility by today's standards, but these articles suggest that some practitioners during the 18th century were interested in using music in medical treatment. At that time, medical care was crude and often dangerous, so a gentle treatment like music therapy was likely welcomed by the public, who often suffered at the hands of the unregulated medical profession (Heller, 1987).

Literature from the 19th Century

During the 19th century, several authors wrote about the use of music to treat physical and mental illness. Articles appeared in music journals, medical journals, psychiatric periodicals, and medical dissertations. Although these reports varied in length as well as quality, they supported the use of therapeutic music as an alternative or a supplement to traditional medical treatment.

The earliest documents produced during this period were dissertations written by two medical students, Edwin Atlee and Samuel Mathews, who attended the University of Pennsylvania. Atlee's work, entitled *"An Inaugural Essay on the Influence of Music in the Cure of Diseases,"* was completed in 1804. He cited literary, medical, and scholarly sources, including material from theorist Jean-Jacques Rousseau, physician and psychiatrist Benjamin Rush, poet John Armstrong, and British musicologist Charles Burney, as well as personal experiences.

The purpose of Atlee's brief dissertation was "to treat the effects produced on the mind by the impression of that certain modification of sound called music, which I hope to prove has a powerful influence upon the mind, and consequently on the body." After defining important terms used in his text, he suggested that music has the ability to arouse and affect a variety of emotions, including joy and grief. The final part of Atlee's dissertation discussed the beneficial effects of music on a variety of mental and physical illnesses and described three cases in which he successfully treated patients with music. In one of his examples, he encouraged a client to resume playing the flute.

Samuel Mathews wrote *"On the Effects of Music in Curing and Palliating Diseases"* in 1806. His dissertation was in some respects similar to Atlee's, but more sophisticated in its use of sources and in the amount of information presented to the reader. Mathews outlined the benefits of music in the treatment of diseases of the mind and body. For example, in order to alleviate depression, he recommended using music that matched the mood of the patient (today this is known as the **iso principle**), because "with the precaution, we may gradually raise the tunes from those we judge proper in the commencement [of a depressed state] to those of a more lively nature." In addition to other citations, Mathews used the Bible to support his assertions, recounting the story of the therapeutic effects of David's harp playing on Saul's psychological difficulties (Gilbert & Heller, 1983).

The dissertations of Atlee and Mathews were strikingly similar in form, content, and physical appearance. Of the many sources they cited, no one person was relied on more heavily than physician/psychiatrist Benjamin Rush, who was a professor at the University of Pennsylvania and a strong advocate for the use of music to treat mental disease. Rush played a major role in creating interest in music therapy during the beginning of the 19th century and likely encouraged Atlee and Mathews to write on the topic (Carlson, Wollock, & Noel, 1981). Their dissertations made a unique contribution to music therapy during the early years of the 19th century.

Music Therapy in 19th-Century Educational Institutions

The use of music therapy in educational institutions for persons with disabilities began in the 19th century. In 1832, the Perkins School for the Blind was founded in Boston by Dr. Samuel Gridley Howe (Heller, 1987). Perhaps due to the urging of his wife, Julia Ward Howe (who composed the lyrics to "Battle Hymn of the Republic"), Dr. Howe, the school's administrator, included music in the curriculum from the beginning. Dr. Howe was instrumental in engaging prominent Boston musicians to help establish ongoing music programs at the school. One of the first of these musicians was Lowell Mason, who taught at the school from 1832–1836. He was responsible for teaching vocal music and piano lessons as well as other music activities. By the time Mason left the school, he had established a strong curriculum in music instruction, which is still in effect today (Darrow & Heller, 1985).

There are other examples of music therapy in institutional settings during the mid-1800s. George Root, a music pupil and friend of Mason, taught at the New York School for the Blind from 1845 to 1850 (Carder, 1972). During the 1840s, William Wolcott Turner and David Ely Bartlett developed a successful music program at the American Asylum for the Deaf, located in Hartford, Connecticut. One student, identified only as Miss Avery, successfully completed a difficult course of piano study. Turner and Bartlett reported her accomplishment in an article entitled "Music Among the Deaf and Dumb," which appeared in the October 1848 issue of the *American Annals of the Deaf and Dumb* (Darrow & Heller, 1985). Music programs were also developed for students with physical disabilities during the early to mid-1800s.

During this period when music therapy was being developed in educational settings, there was also renewed interest in its use as treatment for disease. Three unsigned articles (all entitled "Medical Powers of Music") appeared in *Musical Magazine* within a span of two months in 1840–1841. These reports focused on the linkage of therapeutic music with history, philosophy, and religion, but added little in the way of new information. Much of the material in the articles came from British music historian Charles Burney, whose book *A General History of Music* was published in 1789. One prominent example was the story of King Philip V of Spain, who suffered from depression. During the late 1730s, the famous Italian baroque castrato, Farinelli, who had retired to Spain, was summoned to Madrid to perform for King Philip. It was reported that the king was so moved by the singing of Farinelli that all signs of his chronic depression disappeared, thus assuring the singer lifelong gratitude from the Spanish monarch. The final article in the series cited stories from composers, writers, historians, and performers who had firsthand experience with the therapeutic effects of music (Heller, 1987). Despite their questionable credibility, these articles indicated ongoing interest in music therapy during the first third of the 19th century.

The next substantial support for music therapy was not published until 1874. The article penned by physician James Whittaker entitled "Music as a Medicine," cited an impressive number of American and European sources to support his theory that musical response was linked to physiological, psychological, and sociocultural attributes. Numerous examples were provided to support his belief in the power of music to influence mind and body. Whittaker concluded that the greatest effects from the use of music were on mild forms of mental illness, whereas the treatment of physical ailments and severe mental distress with music was temporary at best.

A second article during that decade was published in the *Virginia Medical Monthly* in 1878. "Music as Mind Medicine" was an edited version of a piece that originally appeared in *The World,* a New York newspaper, dated March 6, 1878. The journal article, edited by Landon B. Edwards, described a series of experiments that took place at Blackwell's Island (now Roosevelt Island), an infamous facility for the care of New York City's indigent, insane citizens. These sessions were held to test "lunatics'" reactions to live music provided by instrumental and vocal soloists. The report began with introductory information about the purpose of the experiment and the people who participated. The principals included the distinguished American pianist John Nelson Pattison, who was credited with initiating the project. Also taking part were New York City Charities Commissioner William Brennan, several physicians, and a number of New York City government officials. A large entourage of musicians also accompanied Pattison and the others to the hospital. The group included 40 members of D. L. Downing's Ninth Regiment Band and several vocalists from the New York Musicians Guild.

The musicians provided music for a large group of patients following a series of nine individual sessions. Pattison directed the individual sessions from a piano, while the doctors assisted by taking physiological data and recording each patient's reaction to the music. The government officials apparently were onlookers, although this was not specifically stated. The article reported that similar sessions had taken place on four previous occasions, about which no information has been located.

The music experiments on Blackwell's Island marked an unprecedented attempt to alleviate the suffering of a large group of persons with mental illness. Authorities who were in a position to implement and maintain such programs, an occurrence not seen previously in the United States, supported the concerts and individual sessions.

During the final decade of the 19th century, two important papers appeared that provided strong support for music therapy in institutional settings and in private practice. In January 1892, George Alder Blumer's treatise titled "Music in Its Relation to the Mind" appeared in the *American Journal of Insanity*. Although the author recognized the therapeutic value of music, he did not support the extravagant claims made by others. Blumer believed that music was a part of moral treatment. The combination of art, reading, music, and physical education provided a well-rounded therapy program for persons with mental illness. Blumer held music in such high regard that he hired immigrant musicians to perform for the patients at Utica State Hospital in New York, where he served as Chief Executive Officer. In fact, Blumer may have been the first person to establish an ongoing music program in an American hospital and should be considered a pioneer in the music therapy movement in the United States.

James Leonard Corning, a prominent neurologist, made another innovative contribution to the advancement of music therapy practice in the late 19th century. His article, titled "The Use of Musical Vibrations Before and During Sleep—Supplementary Employment of Chromatoscopic Figures—A Contribution to the Therapeutics of the Emotions," was published in the *Medical Record* in 1899. Corning's work represented the first controlled attempt to treat mental illness with music. He kept up to date with trends in psychology and neurology and used the information from both professions to fashion his unusual treatment procedures, which he called vibrative medicine.

Using an interesting array of equipment, Corning maintained a consistent environment for testing his patients' reactions to music. He presented music and visual images to his patients as they passed from presleep to sleep. Corning believed that during sleep a person's thought processes became dormant, allowing the penetration of "musical vibrations" into the subconscious mind. Appropriate musical selections (classical music only) helped to transfer those pleasant images and emotions into the waking hours, which suppressed and eventually eliminated the morbid thoughts that plagued his patients. Corning's theories about the relationship between sleep, emotions, and health were based on assumptions that have not been validated by modern research. His work, however, was important, because it represented the first documented attempt to systematically record the effects of music on mental illness.

Throughout the 19th century, music therapy was championed by musicians, physicians, psychiatrists, and other individuals interested in promoting this unique form of therapy. However, these advocates worked independently of each other, so there was little overall growth in its use. During the final decade, articles about music therapy began to appear more frequently in popular and professional journals, and the public began to gain an awareness of the therapeutic possibilities (Davis, 1987). This growth continued into the early years of the 20th century.

Early 20th Century Music Therapy

At the turn of the 20th century, medical care and treatment in the United States was very different than from today. For example, mortality rates were high, many "cures" were still unpleasant, antibiotics had not been developed, and mental illness and intellectual disabilities were largely misunderstood. Hospitals were frequently little more than warehouses designed to isolate but not treat people with disabilities or illness; stays were often long (in 1923 the average hospital stay was almost 13 days compared to today's average of about 4 days). If you were diagnosed with an intellectual or mental disability, hospitalization could last a lifetime. (Starr, 1984; Trent, 1994).

The use of music in hospitals as a soothing, healing, or normalizing agent can be better understood as one considers the extended length of hospitalization and the difficulty of providing care for difficult or "hopeless cases" that did not respond to medical practices of the day. For persons who faced weeks, months, or years of confinement in a hospital or asylum, music was something that could comfort or brighten the mood of patients whose medical needs could be addressed only marginally by typical medical care.

Given the limited understanding of illness and its cause at that time period, it is no surprise that during the first years of the 20th century music therapy gained support only sporadically and its use was often based upon informal observations by physicians or nurses who believed in the potential benefit of music. From time to time, physicians, musicians, psychiatrists, and the general public presented their cases for music therapy in scientific publications, newspapers, and the popular press. Clinical and experimental research provided data to support therapists' contentions that music could be effective in a variety of settings. In addition, a number of short-lived organizations promoted music therapy programs in hospitals, especially for returning World War I and II veterans (Taylor, 1981).

One of the most influential figures to advance the cause of music therapy during the first two decades of the 20th century was Eva Vescelius. She promoted music therapy through numerous publications and the National Therapeutic Society of New York, which she founded in 1903. In *Music and Health*, a publication completed shortly before her death in 1918, she provided a fascinating view of music therapy based on both age-old and contemporary concepts of health and disease. Vescelius felt that the object of music therapy was to return the sick person's discordant vibrations to harmonious ones. She gave precise instructions for the treatment of fevers, insomnia, and other ailments with music.

Perhaps her most unique contribution was the publication of the short-lived journal *Music and Health*. Published in 1913, only three issues appeared in print. Each contained poems and articles by Vescelius and others on the therapeutic applications of music. Additionally, there were advertisements for a course in "musicotherapy" offered by Vescelius. After her death, her sister Louise carried on her work for a short period of time (Davis, 1993).

The first course work in music therapy offered through a university was organized and taught by Margaret Anderton, an English-born pianist who had provided music therapy services to Canadian soldiers suffering from physical and mental disabilities during World War I. During 1919, she taught classes at New York City's Columbia University that prepared musicians for working in hospitals as therapists. She believed that "it is the object of the course to cover the psychophysiological action of music and to provide practical training for therapeutic treatment under medical control" ("Columbia University to Heal Wounded by Music," 1919, p. 59). Like Vescelius, she strongly felt that musicians should be thoroughly trained as therapists before working with patients.

Anderton advocated two principal ways to administer music therapy. For soldiers suffering from psychological conditions, the therapist should provide the music. For those afflicted with physical conditions, the patient should be responsible for producing the music, because it would help to strengthen an injured arm or leg. She also favored the use of woodwind instruments (especially for psychological conditions), because, according to her research, the timbre produced healing effects (Taylor, 1981).

Isa Maud Ilsen, a musician, nurse, and hospital executive, founded the National Association for Music in Hospitals in 1926. Previously, she had served as a teacher of musicotherapy at Columbia University with Margaret Anderton in 1919. She had also been Director of Hospital Music in World War I reconstruction hospitals for the American Red Cross. Ilsen viewed music as a way to alleviate pain for surgical patients and those with physical ailments. Her 20 years as a hospital musician helped to refine her theories concerning music therapy, and, like Eva Vescelius, she viewed a healthy person as one who is in harmony (Ilsen, 1926). Ilsen believed rhythm to be the vital therapeutic component in the music, although she believed that certain styles of music, such as jazz, were inappropriate for treatment.

Like other musicians and physicians during the first half of the 20th century, Ilsen prescribed a specific treatment regimen, using primarily classical music for the relief of a variety of disorders. For severe insomnia, for example, she prescribed a "dose" of Schubert's "Ave Maria." For terminal illness, she believed that Brahms waltzes or

Sousa marches were appropriate. She sometimes used ethnic songs and instrumental music in making her selections, which would suggest some consideration of the patient's music preferences ("Music Prescriptions," 1919, p. 26). Like many other early music therapists, she wanted hospitals to use qualified individuals to administer music therapy programs. Isa Maud Ilsen should be considered an important pioneer in the movement to promote music therapy in American hospitals (Boxberger, 1962).

Like Ilsen and Anderton, Harriet Ayer Seymour worked with World War I veterans as a music therapist, gaining experience and insight into the therapeutic value of music. Inspired by the writings of Eva Vescelius, she published her own guide for the aspiring music therapist in 1920, titled *"What Music Can Do For You."* Over the next 25 years, she actively promoted music therapy through her writings and practical demonstrations. During the Depression of the 1930s, she became involved in the Federal Music Project of the Works Progress Administration, which was an employment program implemented by the Roosevelt administration. Under her guidance, music programs were presented at numerous New York City hospitals and prisons. She conducted experiments to determine the effectiveness of certain types of music on physical and mental disorders (Davis, 1997). Seymour founded the National Foundation for Music Therapy in 1941. As president, she presented lectures and taught classes, emphasizing music therapy techniques used with returning World War II veterans. Her career culminated with the 1944 publication of the first text outlining a course of study in music therapy (Boxberger, 1962; Seymour & Garrett, 1944).

An Instruction Course in the Use of Practice of Musical Therapy presented Seymour's ideas about the appropriate applications of music with a variety of clinical populations. Only brief consideration was given to specific techniques. Essentially, her therapeutic strategy was the same for all clients, consisting of a variety of light classical music selections and folk songs performed by a small group of musicians under the guidance of a lead therapist. According to Seymour, a successful therapeutic experience was achieved through a combination of the music and positive thought, or musical meditation. Because the book was crude in its appearance, organized in a confusing manner, and printed with errors in typesetting and spelling, it is unlikely that the book received wide distribution or use. Despite those shortcomings, Seymour may have used the text to help train some of the 500 music therapy students that she claimed to have worked with between 1941 and 1944 (Davis, 1996).

Although the number of reports of music therapy activity in institutional settings increased dramatically during the first half of the 20th century, music therapy was not widely accepted as a profession by the medical community. The attempts of Vescelius, Ilsen, and Seymour to establish permanent jobs in hospitals, prisons, and schools met with only partial success, probably because of limited support from physicians and hospital administrators (Davis, 1993).

Some doctors, however, actively promoted music therapy. In 1914, Dr. Evan O'Neill Kane, in a letter to the *Journal of the American Medical Association*, enthusiastically endorsed the use of the phonograph in the operating arena for the purposes of distracting and calming patients undergoing surgical procedures. The music was particularly important during the administration of anesthesia, because "the phonograph talks, sings, or plays on, no matter how anxious, busy or abstracted the surgeon, anesthetist and assistants may be, and fills the ears of the perturbed patient with agreeable sounds and his mind with other thoughts than that of his present danger" (Kane, 1914, p. 1829).

In 1915, Dr. W. P. Burdick, who often worked with Kane in the operating room, reported in *The American Yearbook of Anesthesia and Analgesia* that the phonograph was being used not only in operating rooms, but also in wards as a diversion from discomfort and an aid to sleep. Burdick indicated that even the most serious cases improved while the music was playing and that 95% of his patients expressed interest in having the music as part of the healing process (Burdick, 1915).

In 1920, Esther Gatewood further emphasized the use of music in the operating arena, especially during the administration of anesthesia. Like Kane and Burdick, Gatewood advocated patient-preferred music during surgical procedures but believed that it was important to initially match the music to the mood of the client, then to change the temperament of the patient by degree. Gatewood was describing the technique that would later be named the *iso principle*. This principle became more fully developed in the 1940s by Ira Altshuler (Taylor, 1981).

As more reports appeared, the use of music spread from the operating room to other treatment areas. In 1929, Duke University included music for patients not only in operating and recovery areas, but also in both children's and adult wards. Every patient had access to radio reception through earphones or speakers located throughout the hospital. This development represented the first extensive commitment to music therapy by a major American hospital (Taylor, 1981).

In 1930, J. A. McGlinn published an article that reviewed the side effects of anesthesia used in obstetric and gynecological procedures. McGlinn reported that music could effectively reduce patient anxiety during the administration of anesthesia without disrupting operating room routine. Specifically, he recognized four benefits of music, which was chosen to fit the mood of the patient: (1) it effectively masked the sounds in the operating room; (2) it engaged the attention of the patient under local or spinal anesthesia; (3) it relaxed operating room personnel, including nurses, doctors, and other assistants during the surgical procedure; and (4) it provided a source of entertainment for the custodial crews cleaning up after the operation. McGlinn also indicated a bias against jazz and "sentimental" music, believing that it had no place in the hospital (McGlinn, 1930).

Dr. A. F. Erdman continued to champion the cause of music during surgical procedures in the 1930s. Like McGlinn, Erdman believed that music was effective in diverting the patient's attention from the impending operation. Instead of providing music for the entire staff, however, Erdman experimented with a Western Electric music reproducer and earphones, which allowed the patient to hear both the music and instructions from the surgeon. Preferences of the patient were considered when selecting the music prior to the surgical procedure (Erdman, 1934).

Besides its use during surgical procedures, music therapy was also employed in hospital orthopedic and pediatric wards. Harriet Ayer Seymour, founder of the National Foundation for Music Therapy, prescribed specific styles of music for children suffering from ailments such as tuberculosis and physical disabilities. Later, music was successfully used by physicians, including K. L. Pickerall and others, in all phases of a patient's hospital stay, from admission to discharge. In addition to the reduction in anxiety provided by the music, Pickerall noted that medication levels were often reduced and that recovery time was shorter than with clients not receiving music (Taylor, 1981).

Willem Van de Wall was another music therapy innovator noted chiefly for his contributions to the development of music therapy programs in mental hospitals and prisons between World War I and World War II. The Russell Sage Foundation, a philanthropic organization devoted to improving the human condition, provided financial support for his work. Grants led to the publication of a number of important books on music therapy, including a comprehensive work titled *Music in Institutions*, published in 1936.

Van de Wall, like Anderton and Ilsen, lectured on music and health at Columbia University from 1925 to 1932. He also served on the State of Pennsylvania's Bureau of Mental Health, where he was a field representative in charge of music and other therapeutic programs. This position was developed to improve conditions in Pennsylvania mental hospitals (Boxberger, 1963). The first hospital music program developed by Van de Wall in the Commonwealth of Pennsylvania was at Allentown State Hospital for Mental Diseases during the late 1920s.

In 1944, Van de Wall was appointed Chairman of the Committee for the Use of Music in Hospitals, whose purpose was to oversee the progress of music therapy programs in psychiatric hospitals. Boxberger (1963) considered Willem Van de Wall one of the most important 20th-century figures in the development of music therapy in hospitals and institutions.

Ira Altshuler, a contemporary of Van de Wall, was another important individual active in promoting music therapy during the mid 20th century. In 1938, Altshuler initiated one of the first large-scale music therapy programs for persons with mental illness at Detroit's Eloise Hospital. His innovative programs combined psychoanalytic techniques and music therapy methods specifically designed for use with large groups of clients. He later trained some of the first music therapy interns in the United States, working closely with students and faculty from Michigan State University. Dr. Altshuler tirelessly promoted music therapy through numerous publications and presentations and was a founding member of the National Association for Music Therapy in 1950 (Davis, 2003).

Although substantial music therapy activity was recorded during the first four decades of the 20th century, there was no trend toward its regular use. Despite support from such people as Van de Wall, Vescelius, Ilsen, Altshuler, and Seymour, music therapy had still not developed as an organized clinical profession (Boxberger, 1962).

THE DEVELOPMENT OF THE MUSIC THERAPY PROFESSION

In the 1940s, the use of music in the treatment of psychiatric disorders became more widespread, partly due to a gradual change in treatment philosophy. Many therapists, including the eminent psychiatrist Karl Menninger, began to advocate a holistic treatment approach (one that incorporates a variety of treatment modalities). With this shift in philosophy and increased knowledge about its effective applications, music therapy finally became an accepted treatment modality in many hospitals. In addition, the belief that music was somehow "magic" was starting to be dispelled as hospitals and clinics began to sponsor scientific research in music therapy. Much of this effort can be attributed to Frances Paperte, founder of the Music Research Foundation in 1944, and later Director of Applied Music at Walter Reed General Hospital located in Washington, D.C. (Rorke, 1996).

During World War II, numerous organizations, including the Musicians Emergency Fund, the Hospitalized Veterans Music Service, Sigma Alpha Iota, Mu Phi Epsilon, the American Red Cross, and Delta Omicron, provided musicians to Veterans Administration hospitals and later to state institutions. These volunteers assisted hospital staff in organizing ongoing music programs for patients.

By the conclusion of World War II, many United States medical facilities, recognizing the value of music as therapy, employed music programs to assist in the physical and mental rehabilitation of returning soldiers. Although not explicitly referred to as music therapy, the goals were unmistakable: activities implemented by volunteer musicians and music performances by men and women's military bands and choirs were designed as an important element of a wounded soldier's "reconditioning." (Robb, 1999; Sullivan, 2007).

Before the formation of the **National Association for Music Therapy** in 1950, most "music therapists" were unpaid, part-time staff members who worked under the supervision of hospital personnel and who lacked professional status. Many people began to recognize that future growth of the profession would be predicated on effective leadership of trained music therapists. During the 1940s, institutions such as Michigan State University, the University of Kansas, Chicago Musical College, College of the Pacific, and Alverno College started programs to train music therapists at both the undergraduate and graduate levels (Boxberger, 1962). Graduates of these programs comprised the first group of professionally trained music therapists, most of whom worked with persons who were mentally ill.

While music therapy training programs were being developed at a few colleges and universities, movement toward the formation of a national organization was also taking place. The Committee on Music in Therapy of the Music Teachers National Association (MTNA) presented programs during the late 1940s to educate musicians, physicians, psychiatrists, and others in the ways that therapeutic music could be used in schools and hospitals. Ray Green chaired an organizational committee to form a national music therapy association (Boxberger, 1962). The first meeting of the new organization took place in June 1950. Attendees adopted a constitution, set goals, developed membership categories, and appointed a standing committee for research. The National Association for Music Therapy, Inc. (NAMT) was born. The first annual conference was held in conjunction with MTNA in Washington, D.C., during December 1950.

The years following the founding of NAMT focused on improving education and clinical training as well as establishing standards and procedures for the certification of music therapists. Professional publications also enhanced the credibility of the young organization. Monthly newsletters, annual publications, and quarterly periodicals preceded the establishment of the *Journal of Music Therapy (JMT)* in 1964. This journal, edited by William Sears, was (and still is) devoted to research efforts of music therapists.

Probably the most important leader in the field of music therapy during the formative years of NAMT was E. Thayer Gaston (1901–1971). As chairman of the Music Education Department at the University of Kansas, he championed the cause of music therapy during the decades of the 1940s, 50s, and 60s. In collaboration with the renowned Menninger Clinic, a facility in Topeka, Kansas that specialized in the treatment of mental disorders, he established the first internship training site in the United States. In addition, Gaston started the first graduate music therapy program in the United States at the University of Kansas. His "insatiable thirst for knowledge, dedication to scholarship, and unquestioned integrity led to his preeminent position in this field, and many of his associates referred to him as the 'father of music therapy'"(Johnson, 1981, p. 279).

Perhaps the most important action taken by NAMT during its early years was the establishment of the Registered Music Therapist (RMT) credential. This designation was established in 1956 in conjunction with the National Association for Schools of Music (NASM), who served as the accrediting agency. The RMT credential provided assurance to employers that the therapist had met educational and clinical standards set by NAMT and NASM.

As the number of RMTs increased, so did the types of populations served. During the early years of NAMT, music therapists worked primarily with psychiatric patients in large, state-supported institutions. By the mid 1960s, music therapists were also working with adults and children with intellectual disabilities, people with physical disabilities, individuals with sensory impairments. By 1990, music therapy clients included elderly people in nursing homes, patients with medical conditions, and prisoners. During the early years of the 21st century, music therapists continue to work with increasingly diverse clinical populations. In addition to the conditions listed above, significant numbers of music therapists are improving the lives of persons with Rhett's syndrome, AIDS, substance abuse, and terminal illness.

A second organization, the **American Association for Music Therapy**, was established in 1971. Initially called the Urban Federation for Music Therapists (UFMT), the American Association for Music Therapy (AAMT) developed policies and procedures on education, training, and certification that differed from those of NAMT (see Chapter 1). In January of 1998, the National Association for Music Therapy and American Association for Music Therapy merged to create a single organization, the American Music Therapy Association (AMTA).

Since the inception of NAMT in 1950, AAMT in 1971, and AMTA in 1998, the profession of music therapy has continued to grow, with both organizations emphasizing high standards for education, clinical training, and clinical practice. In addition, publications have added to the development of the profession. *Music Therapy*, published annually by AAMT, began in 1980, while a second NAMT periodical, *Music Therapy Perspectives*, began publication in 1984. This semiannual journal provides information on music therapy techniques with specific populations. Since 1998, the *Journal of Music Therapy* and *Music Therapy Perspectives* have served as the two official journals of AMTA. In 1985, a **Board Certification exam** sponsored by both NAMT and AAMT was implemented to strengthen the credibility of the profession. By 2007, more than 2,000 music therapists in the United States worked in diverse settings with a variety of disability groups. The music therapy profession is strong and viable and anticipates continued growth into the 21st century.

SUMMARY

The earliest references to the relationship between music and medicine are found in ancient preliterate cultures. In some of those societies, which exist in parts of the world today, an ill person was seen as a victim of an evil spell, and in others, as a sinner against a tribal god. Music was used extensively in healing rituals by "medicine men," either to appease the gods who had caused the illness or to drive away evil spirits from the patient's body.

Throughout the development of civilization, the relationship between music and healing has complemented the theory of disease prevalent at the time. This evolutionary process has included periods of magic, magico-religious, and rational interpretations of disease. By the 6th century B.C., rational medicine had almost completely replaced magical and religious treatment in Greece. For the first time in history, the study of health and disease was based on empirical evidence. The predominant theory at the time was that of the four cardinal humors developed during the time of Hippocrates.

During the Middle Ages, Christianity influenced attitudes toward sick people, who were viewed as neither inferior nor being punished for their sins. Hospitals were established to provide humanitarian care to persons with physical ailments, although the mentally ill population was still mistreated. The theory of the four cardinal humors was still predominant and provided the basis for the use of music in the treatment of disease.

Advances in anatomy, physiology, and clinical medicine during the Renaissance marked the beginning of the scientific approach to medicine. However, treatment of disease was still based upon the theories of the Greek physicians Galen and Hippocrates. Music was often used in combination with medicine and art to treat medical conditions and also as a preventive measure against mental and physical disorders.

During the Baroque era, the theory of the four cardinal humors continued to dominate but was joined by Kircher's theory of temperaments and affections. Music continued to be closely linked with medical practice. Music was used to treat physical ailments as before but also played an increasing role in the amelioration of mental disorders, such as depression.

Music in the treatment of disease was still popular during the last few decades of the 18th century, but a shift was underway to a more scientific approach to medicine. This change was evident in Europe as well as in the United States. Accounts of music therapy in the United States first appeared during the late 18th century, as various physicians, musicians, and psychiatrists supported its use in the treatment of mental and physical disorders.

During the 19th and the first half of the 20th centuries, music therapy was used regularly in hospitals and other institutions but almost always in conjunction with other therapies. The reports that appeared in books, periodicals, and newspapers persuaded early 20th-century pioneers, such as Vescelius, Anderton, Ilsen, Van de Wall, and Seymour, to promote music therapy through personal crusades and organizations, which were, unfortunately, short-lived. Researchers such as Gatewood, Seymour, and Altshuler attempted to study the reasons why music was effective in the treatment of certain physical and mental disorders; however, their efforts were overshadowed by the lack of trained music therapists and unsubstantiated claims of effectiveness that stunted the growth of the profession until collective research efforts and the establishment of undergraduate and graduate curricula began during the mid 1940s.

During World War II, music therapy was used primarily to boost the morale of returning veterans, but it was also used in the rehabilitation of leisure skills, socialization, and physical and emotional function. Most music therapists during this time served as volunteers under the supervision of doctors and other hospital staff.

With the formation of NAMT in 1950 and AAMT in 1971, professional recognition to the women and men working as music therapists was finally forthcoming. The development of a standardized curriculum, regular publications, an efficient administrative organization, and the merger of NAMT and AAMT to form the American Music Therapy Association in 1998 have all contributed to the growth of the profession. Today, music therapy is recognized as a strong, viable profession.

in-hospital clients (Moe 1998, 2000; Moe *et al.* 2000). There are several studies of the effect of GIM on cancer patients and people living with HIV (Bruscia 1991, 1998a; Bunt *et al.* 2000; Burns 2001). A very important qualitative research project was conducted by Denise Erdonmez Grocke (1999b), whose PhD dissertation 'The Music that Underpins Pivotal Moments in 'Guided Imagery and Music' also contains important information on the history of GIM. Grocke uses a phenomenological method of inquiry, where interviews with seven GIM clients are analysed with focus on pivotal experiences. The music underpinning these experiences is identified and analysed. Other researchers have studied transpersonal experiences and (music) transference in GIM.

Categorization

In Bruscia's systematic account of music therapy models (Bruscia 1998) GIM is placed at the intensive level as a transformative music psychotherapy, because in GIM 'the music experience is therapeutically transformative and complete in, of, and by itself, independent of any insights gained through verbal exchange' (1998, p.219).

3.2 Analytically Oriented Music Therapy – The Priestley Model

Analytically Oriented Music Therapy (AOM) is a further development of what was earlier called Analytical Music Therapy. Together with the tradition of Nordoff-Robbins music therapy, AOM is the most widely used active music therapy form in Denmark. In this music therapy modality the clients are actively involved in clinically organized musical activities and the most applied form of musical performance is improvisation. The improvisations can be tonal as well as atonal and it is the clients' way of expressing themselves in improvising music that influences the form and style of the improvisations. Composed music can also be used, or the activity of composing songs or instrumental music, as a part of the work. In all cases personal and/or functional development of the clients are in focus and not an evaluation of the aesthetic quality of the musical product.

Historical development of Analytical Music Therapy

Analytical Music Therapy (AM) was founded at the start of the 1970s by the English professional violinist Mary Priestley, who undertook her music therapy training at the Guildhall School of Music and Drama in London. As Mary Priestley simultaneously went through a personal psychoanalysis for many years, she began the development of a specific theory to combine music therapy and psychoanalysis. In her clinical work she tried to combine a psychoanalytical and psychotherapeutic understanding of the transference phenomena between the client and the therapist with the understanding of meaning and form of expressions in musical improvisations. Her books and articles mirror this deep interest and her important contribution to music therapy in developing Analytical Music Therapy (Priestley 1975, 1994).

No training programs in the 1970s had this combination. Priestley created a supplementary training module for qualified music therapists that mostly contained a comprehensive ETMT (experiential training of music therapists) module including individual music therapy and *intertherapy* with the students in the client role.

Priestley defines AM as follows:

Analytical Music Therapy is the name that has prevailed for the analytically-informed symbolic use of improvised music by the music therapist and client. It is used as a creative tool with which to explore the client's inner life so as to provide the way forward for growth and greater self-knowledge. (Priestley 1994, p.3)

Clinical applications and applications in training

Mary Priestley primarily developed AM in work with psychiatric clients and in her counselling work with private clients. The application of the method today spreads over a whole range of clients where the symbolic use of improvised music (often combined with fairytales or other stories) in work with ego-weak children and adolescents can also create an indirect way of being better integrated or of helping to get a stronger and more clear self-picture and self-understanding. From being a supplementary training AM has been further developed and expanded to be the foundation and primary theoretical basis for some three- to five-year music therapy training programs. The idea of integrating ETMT into a MA program that is methodologically broad-based (as in Aalborg University) is to give the students basic tools to work psychotherapeutically, with music as the primary tool, with clients with complex psychological problems. The basic idea of ETMT training is to develop the students' sensitivity to a level so that they can better resonate with the problems of their clients.

In work with clients with complex psychological problems this training is important for building up alliances and trustful relationships in the music therapy work. In work with other populations (e.g. multiply disabled children) it is more an underlying source of information for the music therapist to get to know what to do,

when to intervene and how to understand acts and interventions, as these clients cannot feed back verbally.

Analytically Oriented Music Therapy – *a way of understanding and analysing the music*

In Denmark and in Germany the term Analytical Music Therapy is no longer used. It has been replaced by the term Analytically Oriented Music Therapy (AOM) which indicates that the method is no longer solely based on psychoanalytical or analytical psychology theories. It is also based on communication theories, on developmental psychological theories and theories concerning the psychosocial components in developing one's personality. This development can be seen as a logical outcome, where AM is being incorporated as a basic modality in eclectic training programmes.

The understanding and analysis of the music can be examined from different ports of entry.

It is still a tradition that a joint analysis of all three components – the music therapist, the music and the client – is truly emphasized in understanding stages of development in AOM. As an example, Bruscia's Improvisational Assessment Profiles (IAPs) (1987) are often used to focus on certain aspects or stages of development in the musical improvisation most often recorded for both the client and the therapist. Another model for analysing uses Stern's terminology in describing mother/infant communication and can be applied to understand the improvisation (see Chapter 4.1).

As you can see it is important that the music is analysed as a unit that includes both the meaningful expressions of the client and how these expressions influence the simultaneous meaningful expressions of the music therapist. In other words a lot of emphasis is placed on the transference phenomena and the relationship between the therapist and the client.

The session: procedures and techniques

In individual work with AOM the therapist works alone. In group work there are usually two therapists working together adopting roles as primary therapist and co-therapist. In work with clients who are able to verbalize, a session typically starts with the therapist and client intuitively exploring verbally what is meaningful for the client here and now. From there they create a working topic formulated into a playing rule as an inspiration for a musical improvisation. During the music performance this working topic is non-verbally explored and the music therapist can be supportive or creative, improvising in connection with the client's music. He/she can also play a certain role that has been decided beforehand. The music can start tonally or atonally and develop in different directions. It often occurs that a slightly altered state of consciousness arises for the music therapist and client during the act of improvising music. This phenomenon can help in creating new ideas in playing/expressing oneself and in understanding the problem. The responsibility of the therapist is to let go of a full clear awareness of what they want to express, the music in itself can surprise and transform what is being expressed so the transformation in itself becomes a part of the curing elements. Most often the client is prepared for the music performance through one or more centering exercises.

At the same time improvisations are unpredictable. Even if the players start with a certain role that has been decided beforehand. The music can start tonally or atonally and develop in different directions. It often occurs that a slightly altered state of consciousness arises for the music therapist and client during the act of improvising music. This phenomenon can help in creating new ideas in playing/expressing oneself and in understanding the problem. The responsibility of the therapist is to let go of a full 'controlling' consciousness, while at the same time keeping an overview of the total act of interplaying. So the therapist is working in a condition of double awareness. Strong feelings expressed musically by the client must be supported or contained by the therapist. Finally, the client may want to play alone and to 'be listened to carefully' by the therapist. The client may also, on occasions, want the therapist to play some familiar classical music or 'caring' improvised music for him/her while he/she is listening and 'being nursed' by the music. A basic problem for psychiatric clients is to be motivated to stay engaged in an activity in order to give a chance for a process to be developed. Overall AOM is a caring method but the work can develop into a strong emotional confrontation for the client, when sufficient trust is present between the therapist and client.

It is important in AOM work that there is a verbal reflection after the playing activity in order to give the clients the possibility to make conscious the inner movements that were provoked in the musical improvisation. Normally a session ends with a final music improvisation where the material that was brought up in the session is digested as much as possible.

Philosophically AOM can de defined as music in therapy, as the music is used to symbolically express inner moods, emotions and associations. It often happens though, that the music 'takes over' and starts to live a life of its own during an improvisation, so that new and unexpected sounds, tones, rhythms and melodies turn the client and therapist away from the intentions they began with. Therefore one can say that AOM often becomes both *music in therapy* and *music as therapy* simultaneously. It is therefore important that the music therapist is flexible in his/her abilities of playing the piano, the drum-set and also other instruments.

PLAYING RULES

To find a focus or a topic for the musical improvisation is called: (creating) playing rules. Many different categories of playing rules can be created and used according to the type of problem. In the form of short-term AOM, there is often both an overall working topic for the course of treatment, and also playing rules created in each individual session. The task of the playing rules is to make the client connect with, and express, certain emotions, fantasies, dreams, body experiences, memories or situations through the music. The playing rule acts as an inspiration and a basis for the inner imagination and for the emotional and sensational experiences that emerge during music improvisation.

At the same time improvisations are unpredictable. Even if the players start with a clear awareness of what they want to express, the music in itself can surprise and transform what is being expressed so the transformation in itself becomes a part of the curing elements. Most often the client is prepared for the music performance through one or more centering exercises.

Documentation

Priestley herself has written two books on techniques and theory creatively exemplified with a lot of clinical narratives (Priestley 1975, 1994) and she has also edited a range of articles on certain aspects of AOM. AOM today extends all over Europe and influences a number of training programs. Over the last 10 years interest in AM has been growing in the USA, where Scheiby today offers a supplementary training module based on Priestley's original AM model for trained music therapists. AOM is still identified by most music therapists primarily as a method for work with psychiatric clients and for counselling work. In Denmark two books in a series of books on music therapy in psychiatry (Lindvang 2000; Pedersen 1998) are based on AOM, and many chapters of international books (Bruscia 1998a; Wigram and De Backer 1999b) also are AOM oriented.

In AOM work with all kinds of client populations, the focus is on the client's self-healing forces and mental resources. In work with multiply disabled clients who cannot verbalize, AOM can be used to gain contact and to communicate at a very basic level. Understanding of stages of development has to be based on the condition of the clients and often has to be seen as tiny nuances.

AM practice has given much inspiration to qualitative research projects, where PhD projects by Langenberg (1988), Mahns (1997), Hannibal (2001) and Hadley (1998) should be mentioned. Bruscia (1998, p.219) places AM (here AOM) on the advanced level of treatment as this therapeutic modality aims at letting the client obtain deep insight, integration and transformation of complex psychological problems.

3.3 Creative Music Therapy – The Nordoff-Robbins Model

Paul Nordoff, an American composer and pianist, and Clive Robbins, a British trained special educator, collaborated together to pioneer one of the most famous improvisational models of music therapy developed over the last 50 years. This approach is now called Creative Music Therapy, and is known worldwide as the Nordoff-Robbins approach.

The history of Creative Music Therapy

Their method, developed between 1959 and 1976, has been taught in several countries, including Great Britain, Germany, USA, Australia, Japan, South Africa, Canada and Norway, and students of this method tend to continue using the approach in their clinical work. Most of the early development of Creative Music Therapy was aimed at children with learning disabilities, from the mild end of the spectrum to the severe, including Down's syndrome, emotionally and behaviourally disturbed, mentally and physically handicapped and children with autism. Paul Nordoff died in 1976, and Clive Robbins then further developed his work with Carol Robbins, his wife, introducing a new focus with hearing-impaired children, while maintaining the application of this model to mainly handicapped and emotionally disturbed children.

Philosophical orientation

In their early years of developing a music therapy method, Nordoff and Robbins were influenced by the ideas of Rudolf Steiner and the anthroposophic movement in humanistic psychology. Here they developed the idea that within every human being there is an innate responsiveness to music, and within every personality one can 'reach' a 'music child' or 'music person'. This idea was very important in their work with the handicapped population, where despite severe degrees of learning disability, and often severe physical disability, they believed in the potentially normal and natural responsiveness to music, and the power of music to enable self-expression and communication. Later, Robbins and Robbins related their therapeutic goals to the humanistic concepts of Abraham Maslow, including in their framework the aspiration towards self-actualization, peak experiences and developing special creative talents. Their relationship with the client is built on a warm, friendly approach, accepting the child as he/she is, recognizing, reflecting and respecting the child's feelings, allowing the child choice, and a non-directive approach to give the child autonomy, and the therapist the role of following and facilitating.

Method and approach in sessions

The Nordoff-Robbins style of work is unique and often easily recognizable. To begin with, it involves placing music at the centre of the experience, and musical responses provide the primary material for analysis and interpretation. They argue the need for highly skilled musicians and, as the use of a harmony instrument is central to their working style, they have predominantly trained therapists in the sophisticated use of piano (and in rare cases, guitar) in improvised music making. In individual therapy, the clients are typically offered a limited channel for their musical material, mainly the cymbal and drum, together with a strong encouragement to use their voice. In group work other instruments are involved – pitched percussion, reed horns, wind instruments and various string instruments.

In much of the individual work, and very much as a mark of their style of therapy, Nordoff-Robbins (where at all possible) work in pairs. One person establishes a musical relationship from the piano, while the other therapist facilitates the child's responses and engagement. This idea is based on the model of work employed with Paul Nordoff as the pianist/therapist, and Clive Robbins as the other therapist. Paul Nordoff's own music, mostly tonal in style, has also formed one of the foundations of the musical engagement, by which we mean that he developed a unique style of improvising that is evident in their two books of playsongs for children. Mostly, the therapists use creative improvisation, and create an engaging musical atmosphere from the moment the client enters the room to the moment he/she leaves.

The style of work, and their approach, comes within the conceptual framework of *Music as Therapy*, where the music provides the therapeutic catalyst through which change will take place. Music occurs almost throughout the session, and the therapeutic relationship is formed *in* the music. The therapists work through phases in their therapy:

- Meet the child musically...Evoke musical response...develop musical skills, expressive freedom, and interresponsiveness... (Bruscia 1987, p.45).

Clinical applications and process of therapy

Nordoff and Robbins offered a significant perspective on how music can be used in music therapy. The improvisational style must be free from musical conventions, and flexible. Intervals are important and represent different feelings, when used in melody. Triads and chords can be used in special ways – for example, the tonic triad to indicate stability, while inverted triads represent dynamic movement. Improvised music should also include musical archetypes, such as *organum*, exotic scales (Japanese, Middle Eastern), Spanish idioms and modal frameworks.

Music-making is the primary focus of therapy sessions, and from the early development of individual therapy, the experience of music was all-pervasive during the session. When working with children, clients are frequently brought into the therapy room while a welcoming music is being played by one therapist on the piano, and at the end of the session, they go out of the room to music. Music in the form of 'clinical improvisation' is used to establish a relationship with the client, provide a means of communication and self-expression, and effect change and the realization of potential. It is the belief in music itself as the medium of growth and development that is at the core of this approach, and the belief that in each person, regardless of disability, ill health, disturbance or trauma, there is a part which can be reached through music and called into responsiveness, thereby enabling healing and the subsequent generalization into the client's life (Etkin 1999).

The therapists often provide a musical frame, frequently establishing clear rhythm and pulse, and particularly, singing about what a patient is doing while they are doing it in order to bring into focus the experience that is occurring. Any musical expressions produced by the client, vocal or instrumental, are incorporated into a frame, and encouraged. The skill of the therapist is brought into play in providing an appropriate musical frame or context for the client's expressions, matching, mirroring or reflecting their musical material. The therapist pays close attention to responding musically to the quality, timbre, pitch, dynamics and inflection of the client's vocal, instrumental and body expression.

The clinical application of Creative Music Therapy has been introduced in a wide-ranging and diverse way. The graduates of the courses in the Nordoff-Robbins method, particularly in New York, London, Sydney, Pretoria and Witten/Herdecke in Germany, have diversified the approach to work with adult patients in the areas of neurology, psychiatry and terminal illness. The method has been tremendously developed through research and extension of applications (Aigen 1991, 1996, 1998; Ansdell 1995, 1996, 1997; Brown 1999b; Lee 1996, 2000; Pavlicevic 1995, 1997; Pavlicevic and Trevarthen 1994; Streeter 1999a; Neugebauer and Aldridge 1998).

Documentation

This model of music therapy has also developed methods of analysing what is going on and how the therapy is progressing. A number of scales have been generated including:

- thirteen categories of response
- child-therapist relationship
- musical communicativeness
- musical response scales: instrumental rhythmic responses, singing responses.

Case studies are the most typical way by which therapists working in the Nordoff-Robbins tradition document their work. The material is often presented as a 'story', a narrative description of the process and progress of therapy. Howat (1995) presented a lengthy and detailed account of individual work with a young 10-year-old girl with autism called Elizabeth, documenting more than 100 sessions over a five-year period. The narrative descriptions, sometimes brief and sometimes more detailed, mainly focused on her musical behaviour in the sessions, explaining how she played with many examples and interpretations of the emotional expression present in her playing. Life events were also included in order to make a context for the musical process in the therapy.

Etkin (1999) described a period of therapy with an emotionally, physically and socially abused and deprived child called Danu. She described the way that Danu played during the initial assessment session, and then set out the case study in the stages of therapy: early work – emergence of songs and stories – disclosure – endings. A method of improvisation called 'singspiel' or 'sprechgesang' featured strongly in the therapy sessions, and there was significantly more verbal material than is typical in other case studies from this tradition. From this example, it is clear that while the original conceptual model of Nordoff-Robbins provides the basis for a strong and grounded training, individual therapists develop methods and techniques out of their primary approach. Piano-based improvisation still forms the foundation, but guitar-based improvisation developed by Dan Gormley in the USA, jazz and blues improvisation styles more culturally effective with some populations in New York developed by Alan Turry, and Aesthetic Music Therapy recently defined by Colin Lee in Canada, amongst others, have emerged from the initial foundations of Creative Music Therapy.

Because Paul Nordoff and Clive Robbins lived in Denmark for a period of time, and also taught in Norway, there are therapists in both countries that follow their style of work, and others who incorporate their concepts at a more general level with certain clinical populations. Claus Bang, the Danish music therapist and audio-speech therapist, has translated the playsongs for use in Denmark.

Creative Music Therapy has lasted the test of time, and is a much practised model of music therapy, more now than ever before, as can be seen in the increasing number of case studies using this approach in the music therapy literature. It is relevant to look first at the writings of Paul Nordoff, and Clive and Carol Robbins (1971, 1977, 1980, 1998). Many other examples of case material can be found in the many anthologies and books, including amongst others *Clinical Applications of Music Therapy in Developmental Disability, Paediatrics and Neurology* (Wigram and De Backer 1999a), *Case Studies in Music Therapy* (Bruscia 1991), *Music for Life* (Ansdell 1995), *Music Therapy in Context* (Pavlicevic 1997) and *Being in Music: Foundations of Nordoff-Robbins Music Therapy* (Aigen 1996).

3.4 Free Improvisation Therapy – The Alvin Model

Juliette Alvin was a pioneer of music therapy and developed a foundation model for improvisational music therapy between 1950 and 1980. She was an internationally famous concert cellist, studying with Casals, and she strongly believed in the effect of music as a therapeutic medium. Her own definition of music therapy was:

> The controlled use of music in the treatment, rehabilitation, education and training of adults and children suffering from physical, mental or emotional disorders. (1975)

She argued that the analytical concepts of Freud underpin the development of music therapy, as music has the power to reveal aspects of the unconscious. While not requiring one to be 'Freudian' to believe in this important concept, Alvin's theory was built on the primary statement: *Music is a creation of man, and therefore man can see himself in the music he creates.* This idea was developed alongside Alvin's perception of music as a potential space for free expression. She cites Stravinsky as one of the single most important influences on music in the twentieth century, because his compositions broke the 'musical rules' in terms of harmony, melody, rhythm and form, and allowed us to make and experience a range of dissonant and atonal sounds that had previously been taboo. This opened the door for her development of free improvisation therapy, where clients and therapists can improvise without musical rules, and where the music can be an expression of the person's character and personality through which therapeutic issues can be addressed.

SESSION FORMAT
Alvin's method is musical:

- All the client's therapeutic work centres around listening to or making music.
- Every conceivable kind of musical activity can be used.
- Improvisation is used in a totally free way, using sounds or music that are not composed or written beforehand.
- By sounding the instruments in different ways, or by using unorganized vocal sounds, inventing musical themes allows great freedom.
- Free improvisation requires no musical ability or training, and is not evaluated according to musical criteria.
- The therapist imposes no musical rules, restrictions, directions or guidelines when improvising, unless requested by the client. The client is free to establish, or not establish, a pulse, metre, rhythmic pattern, scale, tonal centre, melodic theme or harmonic frame.

These were revolutionary concepts for music therapy in the 1960s, as the main schools in the USA used conventional, precomposed music in more behaviourally orientated therapy. Only Paul Nordoff and Clive Robbins' approach in Creative Music Therapy came close to this, although their music was more conventional and structured, and analytical theory was not inherent in their method.

The History of Free Improvisation Therapy

In 1959, Alvin founded the British Society for Music Therapy, and subsequently founded the post-graduate course of music therapy at the Guildhall School of Music and Drama in London in 1968. During her extensive travels in Europe, the USA and Japan, she was invited to start a course in music therapy at university level, but she firmly believed in the importance of highly trained and experienced musicians as potential music therapists, so she chose to begin her course at a music conservatoire, where the emphasis was on musical training and skill, rather than academic knowledge. The programme at the Guildhall School of Music is still running (now in collaboration with the University of York, which validates it), and graduates from her courses, including Tony Wigram (Denmark), Leslie Bunt (United Kingdom) and Helen Odell-Miller (United Kingdom), have furthered her methods in other training courses. Mary Priestley (United Kingdom) also trained with Juliette Alvin, and went on to found Analytical Music Therapy. Many therapists and teachers of music therapy have been influenced by her methods, and her model of Free Improvisation Therapy is still taught and used in clinical practice (Alvin 1975, 1976, 1978).

CLINICAL AREA

Alvin worked in psychiatry, and also focused her work on children, including those who are autistic, mentally handicapped, maladjusted, and physically handicapped.

Alvin taught about the importance of developing the client's relationship with music. In her work with people with autism and developmental disability in particular, she proposed that the client's relationship with the instrument was the primary and initial therapeutic relationship. The musical instrument, for Alvin, can be the container of the negative feelings projected by the client, and represents a 'safe intermediary object'. After this, clients become attracted to, and form relationships with, the instrument of the therapist, centring their feelings in the music created together. It is after going through this process that the development of a relationship directly between client and therapist occurs. So her concepts relating to the objectives of the therapy, the process of the therapy and the successful outcome of the therapy start and develop in the musical relationship. This was a seminal and unique contribution to the theory in the musical relationship.

From a psychotherapeutic and theoretical point of view, Alvin worked within the concept of an 'equal term relationship' where the therapist and client share musical experiences at the same level, and have equal control over the musical situation. This is very significant as a concept, and explains much about the remarkable effect of her therapeutic approach, and her success in drawing out the potentials and strengths of clients with whom she worked. Autistic, maladjusted and physically handicapped children responded eagerly to her approach, when she would offer them an empathic and sensitive musical frame.

Clinical applications: approach and method

Alvin proposed the potential to use different approaches in different situations, and this 'eclectic model' has caused some controversy. She mainly worked from a humanistic and developmental point of view, often describing in her many cases, changes in the client's behaviour that represented underlying changes in their capacities. When working in the field of psychiatry, she approached clients from a more analytical perspective.

Alvin believed the therapist's instrument was his/her primary means of communication and interaction. She herself used a method of 'empathic improvisation' when she used her cello. This involved gaining an insight and understanding about a client's way of being, mood and personality, and then reflecting it back through improvised playing on her cello. This was 'playing for the client' and therefore receptive in style. The therapists can also, through this method, introduce themselves to the client in a safe and non-threatening way, adjusting their playing to the listening responses of the client.

Documentation

Alvin wrote extensively about her concepts and ideas of music therapy in her main books and many clinical articles. Her books were:

- *Music Therapy* (1975)
- *Music Therapy for the Handicapped Child* (1976)
- *Music Therapy for the Autistic Child* (1978).

Probably the most useful overview of Alvin's theory, method, clinical approach and methods of assessment and evaluation can be found in Unit 3 (Chapter 3) of Bruscia's *Improvisational Models of Music Therapy* (1987). As well as what we know and understand about the psychotherapeutic functions of music in music therapy, Alvin placed emphasis on the importance of understanding the physiological effects. She said one needs to link the psychological effect of music with the physical effect, and used the examples of shamans and witch doctors from primitive cultures to illustrate this idea. Music therapists need to understand human physiology and the way the body reacts to music and sound to fully understand the influence of music within music therapy. Alvin has defined some important concepts for our understanding of music therapy within Free Improvisation Therapy:

- analytical concepts of music
- psychological functions of music
- physiological functions of music
- functions of music in group music therapy.

She formulated a descriptive approach to evaluating the effects of music and music therapy, including evaluating listening responses, instrumental responses and vocal responses.

Alvin's method, and her concept of the role of the music therapist, places the level of therapy at either augmentative or primary. In her own clinical work, Alvin worked as part of multidisciplinary teams in hospitals and units, but also with individual clients in private practice – as a primary therapist.

Alvin died in 1982, as music therapy in Great Britain was becoming a regulated profession within the health and social system. Her contribution both at a theoretical and clinical level was foundational in promoting the value of music therapy, and in beginning a course in the United Kingdom that placed music skill and competence at the centre of music therapy training and clinical practice. Alvin holds a place in history as one of the earliest and perhaps most eclectic and inspirational pioneers.

3.5 Behavioural Music Therapy

Behavioural Music Therapy (BMT) was mainly developed in the USA and still forms a primary method of intervention. It is a method that is defined as follows:

... the use of music as a contingent reinforcement or stimulus cue to increase or modify adaptive behaviours and extinguish maladaptive behaviours. (Bruscia 1998)

Dr Clifford Madsen, together with Dr Vance Cutter, published an article in 1966 describing BMT. It is a form of cognitive behaviour modification, and involves applied behaviour analysis. The main concepts are that music is used in treatment in the following ways:

- as a cue
- as a time structure and body movement structure
- as a focus of attention
- as a reward.

As in behaviour therapy, the focus of the treatment is towards the modification of behaviour – conditioning behaviour – which can be measured by applied behaviour analysis. Whether one is working with an autistic child or a depressed adult, the process involves the concept of stimulus-response, and music is used to change behaviour, and reduce symptoms of a pathology, rather than an attempt to explore the cause of behaviour. Research has been undertaken in the USA, and many of the studies have used recorded music in order that the studies can be replicated. Rigorous scientific standards are applied to ensure that the effects of BMT are accepted and recognized in the scientific community. The term behaviour is understood as an all-inclusive concept. Therefore the object of the therapy is the control and manipulation of many different types of behaviour including:

- physiological behaviour
- motor behaviour
- psychological behaviour
- emotional behaviour
- cognitive behaviour
- perceptual behaviour
- autonomic behaviour.

In practice, music in any form is used in conjunction with a behaviour modification program. An example of this would be a program to encourage a person with learning disability to increase their attention span in a group. If the person sustains their attention on a task, or on the therapist, they will receive music (perhaps in the form of songs or a played piece). If their behaviour deteriorates, music is withdrawn. Based on the assumption that the person wants music experiences, the program is designed to increase his/her attention because of his/her motivation to get musical experiences.

Another example is the research and clinical work from Professor Jayne Standley of Florida State University who measured the effect of music in reinforcing sucking behaviours in premature infants. When the babies sucked, they received musical stimulation, and when they stopped sucking, the music was withdrawn. The results of Standley's research showed that the introduction of the musical stimuli resulted in increased sucking, and improvements in weight gain and health leading to earlier discharge from the intensive care unit. Standley also investigated the music that was effective as a stimulus (Standley 1991b, 1995, 1998).

Participation in musical activities is also used in BMT. Whether one is working with geriatric patients, psychiatric patients or adolescents with developmental disability, the structuring and implementation of musical activities such as singing, music and movement, playing and dancing are used to encourage non-musical goals and objectives such as:

- social engagement
- physical activity
- communication
- cognitive processes
- attention and concentration
- enjoyment
- reduction and elimination of antisocial behaviour
- independence skills.

Research and documentation

The music that is used in BMT varies widely, and some research has been done to define what type of music will promote the achievement of therapeutic and treatment objectives. For example, pulsed, rhythmical music is used with patients with Parkinson's disease to promote good walking patterns (Thaut 1985). Old songs, familiar melodies and hymns are used with geriatric patients to promote attention, engagement and memory. Also, music with slower tempos such as largo, adagio and andante are used when attempting movement or dancing with older adults. With patients with senile dementia, short songs and pieces are recommended to cope with short attention span.

Applied behaviour analysis in BMT allows the therapist to measure over time the effect of music therapy intervention. If a child is acting out in school, the introduction of guitar-based music therapy sessions may give the child motivation and stronger self-esteem. Applied behaviour analysis can measure the number of defined, asocial behaviours that are targeted during periods when the child receives music therapy, and also when music therapy is withdrawn for a period. Using reversal designs and multiple baselines, the therapist and researcher can evaluate the actual efficacy of music therapy intervention over time, by comparing the number of behavioural incidents that occur during the different periods of intervention with periods of non-intervention.

BMT is a good example of music *in* therapy, as the role of music is to act as a stimulus and reinforcer of non-musical behaviour. Although the therapist may be interested in the patient's way of making music, and their expression and communication through music, the main focus for therapy and evaluation is to achieve changes in the client's general behaviour.

Bruscia (1998, p.184) places BMT at the augmentative level, because this therapeutic direction works with limits and goals that specifically address symptoms, and, to a lesser degree, with the client's personality or general development.

CHAPTER 4

Basic Therapeutic Methods and Skills

There are many different therapeutic methods that are applied in music therapy when using improvisation. Bruscia (1987, p.533) began with a description of 64 'clinical techniques' and with the increasing volume of published literature on music therapy over the last 12 years, further techniques and methods used in therapy have been reported (Codding 2000, 2002; Pedersen 2002; Staum 2000; Wigram and Bonde 2002; Wigram and De Backer 1999a, 1999b; Wigram, Pedersen and Bonde 2002).

Therapy methods can either be used intentionally (or intuitively) in therapy work with clients or they can be the objects of analysis when reflecting on a period of free-flowing improvisation to explore what was actually happening. It is not usual for music therapists to pre-plan exactly the method they might use, unless they are working in an activity-based model, or with a structured assessment procedure. In improvisational music therapy, particularly, the model requires an adaptive and flexible response to the way the client begins to make music. There can be a certain degree of planning based on the assessment that has taken place and an estimation of the client's needs and the objectives of therapy that will promote certain techniques above others. However, it is more typical that improvisational music making occurs, and within that music making intuitive judgements about therapeutic method are made based on the 'here and now' experience. Music therapists don't remain exclusively attached to one musical technique or therapeutic method for a set period of time, and might fluctuate between a number of different methods (as well as musical techniques) over the course of a single improvisation.

This chapter presents, discusses and exemplifies certain specific methods that are commonly used in music therapy, in order to provide methods within which the musical techniques that have been described in the previous chapter can be applied.

81

It is very useful to practise these techniques together with another person, first of all playing the experience and subsequently responding to the musical production of another. Each technique will include a musical illustration, complemented by an example on the CD.

4.1 Mirroring, imitating and copying

Mirroring and *imitating* are frequently used as empathic techniques where the music therapist intends to give a message to the client that they are meeting them exactly at their level and attempting to achieve synchronicity with the client. Bruscia has described the technique of mirroring as 'synchronising' – doing what the client is doing at the same time'. I define mirroring in a similar way but with a slightly broader explanation, in order to suggest to clinical practice that mirroring involves more than just musical behaviour:

> *Mirroring: Doing exactly what the client is doing musically, expressively and through body language at the same time as the client is doing it. The client will then see his or her own behaviour in the therapist's behaviour.*

This can only be achieved musically, where the client's music is both simple enough and predictable enough for the therapist to anticipate how to mirror exactly what the client is doing. This also applies to the physical behaviour of the client. In order for the mirror to be exact, the therapist may also need to pay attention to using a very similar instrument as the client in order to achieve a mirrored response. However, it is possible to accomplish mirroring by using a different instrument. Example 17 on the CD gives an illustration where the therapist can use the piano almost as a drum while the client plays on a drum.

CD17: Mirroring – client on drum + therapist on piano

'Close enough' mirroring is a technique where the therapist is doing almost exactly the same as the client but due to technical reasons cannot copy exactly. For example, this would work very well where the client is randomly playing notes on a metallophone and the therapist mirrors that by playing as near an imitation as possible at the same time, achieving the direction of the melody and the general contour of the melody without necessarily matching exact notes.

Conceptually, we can see the identities of the participants in mirroring (client and therapist) in a very symbiotic relationship, where they become fused and undivided. Figure 4.1 illustrates the place of the therapist and client inside two circles where the integration of one circle into another represents the closeness of the material.

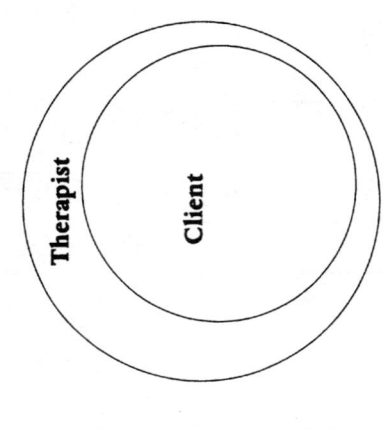

Figure 4.1: Musical closeness in mirroring

Imitating or copying are also empathic methods of improvisation and imitating has been defined by Bruscia as 'echoing or reproducing a client's response after the response has been completed'. This relies on the client leaving spaces in the music for the therapist to imitate what he or she is doing. It should be used quite specifically, and caution needs to be exercised as imitating or copying a client's production might appear as though you were either teasing or patronizing the client. While mirroring and copying are relatively simple methods, they can also be quite confronting to a client, and can be risks, for example, with clients with paranoia or thought disorder for whom this method may excite irrational fears. This approach needs to be used sensitively and appropriately. Nevertheless, it is a therapeutic strategy to help a client to be aware that musically you are echoing and confirming what they have done.

4.2 Matching

I regard *matching* is one of the most valuable of all the improvisational methods that can be applied in therapy. It is, in my approach, a typical starting point to work together with the client musically, from which a number of other therapeutic strategies or methods emerge. It is also an empathic method, as the music produced by the therapist in response to the client confirms and validates their playing and their emotional expression.

I have defined the term to be quite inclusive:

IMPROVISATION

Matching: Improvising music that is compatible, matches or fits in with the client's style of playing while maintaining the same tempo, dynamic, texture, quality and complexity of other musical elements (Wigram 1999a).

To achieve a 'match' in musical terms means that the therapist's music is not identical to the client's, but is the same in style and quality. Therefore the client experiences that the therapist's music 'fits together and matches' his or her own production.

Conceptually, we can begin to see the two separate identities of the participants (client and therapist) in their musical relationship, where they are together, congruent and matched musically, but with some individual differences that show emerging separateness. Figure 4.2 shows two circles separating, representing the matched material but separating identities of the therapist and client.

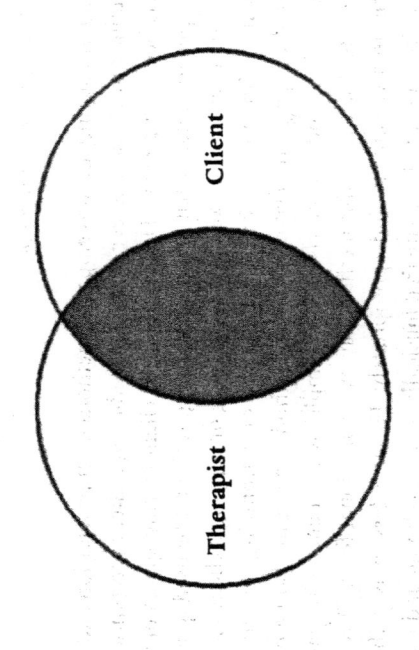

Figure 4.2: Musical connections in matching

Bruscia does not include matching as a term, but incorporates the idea into a definition of reflecting. Pavlicevic (1997) has referred to it in her book *Music Therapy in Context* giving a different conceptual understanding. She thinks of matching as 'partial mirroring where, for example, the client plays a definite and predictable musical pattern, and the therapist mirrors some, but not all, of the rhythmic components' (p.126).

My experience and use of matching in therapy is more as an equal, complementary style of playing together, as illustrated in Figures 4.3, 4.4 and 4.5, and demonstrated in CD18, CD19 and CD20. The CD examples start with the 'client' playing, and show how the therapist joins in, matching the music of the client. In the first examples (Figure 4.3 and CD18) the rhythmic style of the client is revealed as short, quite stable rhythmic patterns in a regular pulse. As the improvisation develops, the

BASIC THERAPEUTIC METHODS AND SKILLS

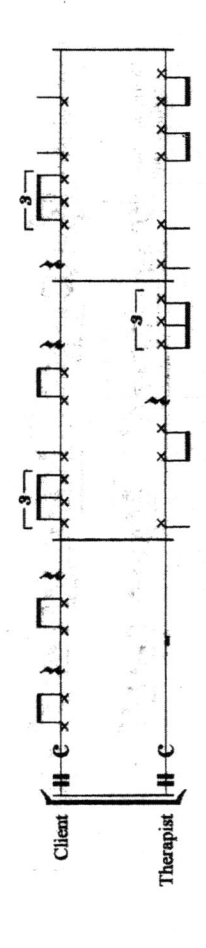

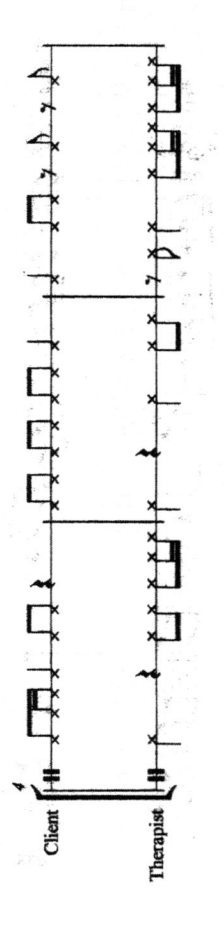

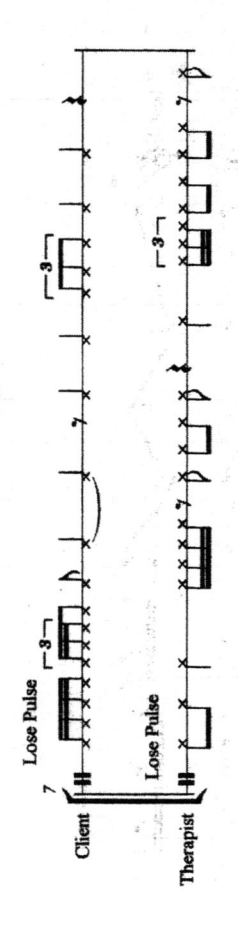

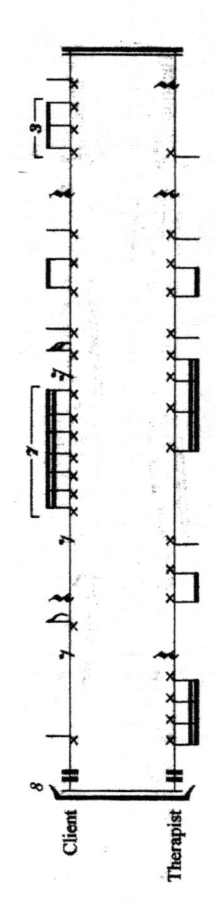

Figure 4.3: Matching: client on bongos, therapist on conga

Figure 4.4: Matching: client on xylophone, therapist on metallophone

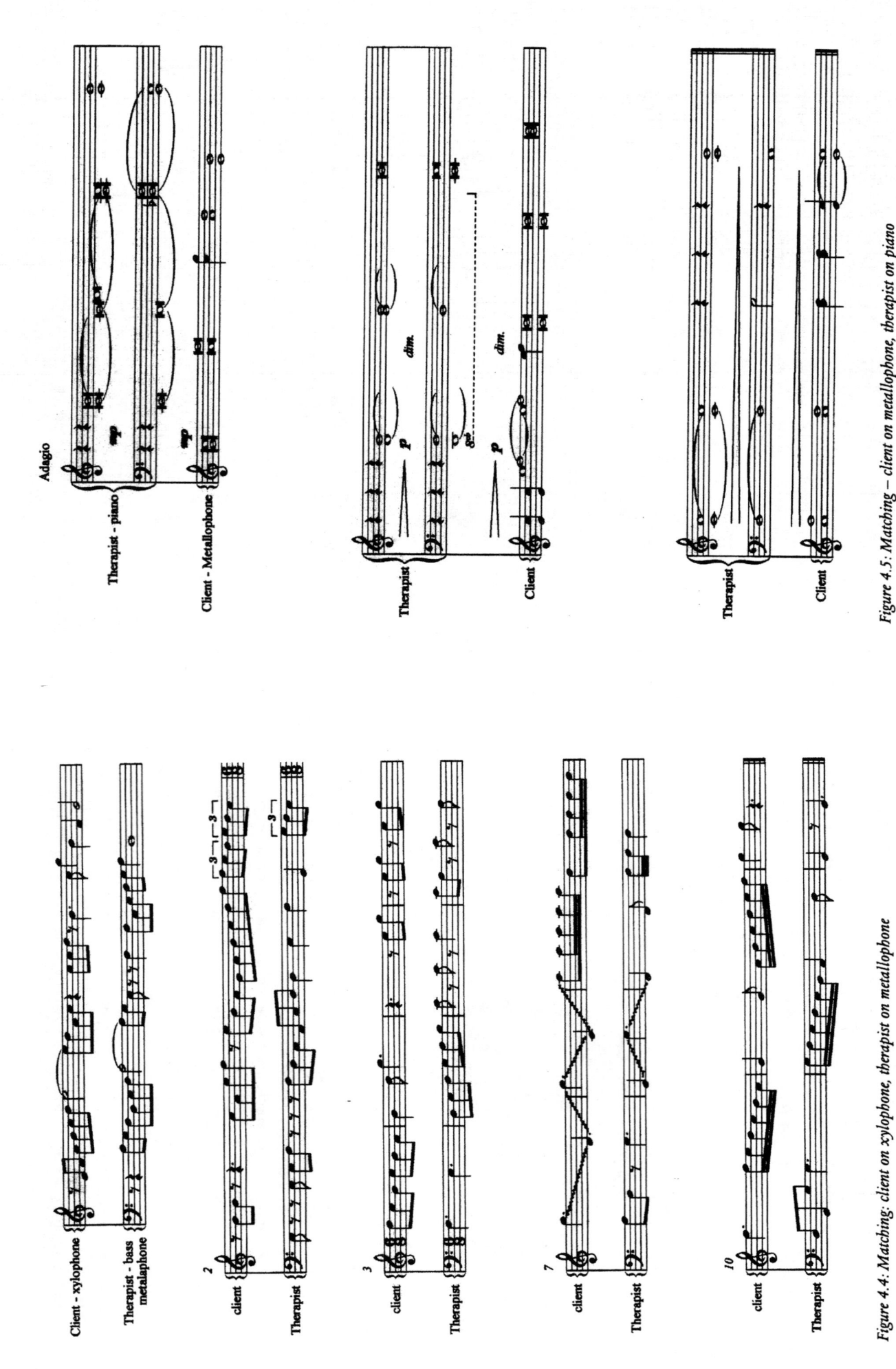

Figure 4.5: Matching – client on metallophone, therapist on piano

style changes with a loss of any sense of pulse in the client's playing, and the therapist can be heard to adapt and sustain matching.

CD Example 18: Matching – client on bongos, therapist on djembe

In the next examples (Figure 4.4 and CD19) melodic matching is illustrated. Here the emphasis is on style of the melody, in particular phrase lengths, step-wise or large interval movement and tonality. The client's material changes as the example goes on, and the therapist can be heard to adapt to this change.

CD Example 19: Matching – client and therapist on melodic instruments

Finally, Figure 4.5, CD 20 gives an example where the therapist (piano) uses chords to match with a client (metallophone) who is playing sustained, two-tone sounds, without any sense of rhythmic or harmonic direction. In the therapeutic process of matching it is very important to stay true to the client's music, and not attempt to modify, change or manipulate. At this stage of therapeutic intervention, using the matching method, therapeutic directions or 'solutions' are not the primary objective, and may emerge later. The engagement, close to the tradition and goal of client centred therapy, is to offer 'unconditional positive regard' in the form of acceptance and matching.

CD Example 20: Client on xylophone, therapist on piano

Matching exercises

The CD has two examples of a person playing that provide an opportunity to practise the therapeutic method of matching. The first part of the process in matching is to listen to and analyse the musical components of a client's production, also taking into account their level of expression in their body and their face. However, as these examples are presented on CD, the latter information is not available and one needs solely to consider the musical elements.

Table 4.1 identifies the musical elements for these two examples in order to clarify the type of music the therapist should produce to match and empathize with the client's material.

Table 4.1 Structured matching exercises

Example	Style	Rhythm	Dynamic	Tonality
1 (CD21)	Folk	4/4 regular	Soft and slow	Pentatonic
2 (CD22)	Jazzy	Irregular	Wide range	Atonal

4.3 Empathic improvisation and reflecting

Mirroring, copying and *matching* involve a more technical exercise of creating a musically congruent response to the client, attending primarily to the balance and salience of musical elements, as well as body language and expression. *Empathic improvisation* and *reflecting* require a response that is more specifically connected to the emotional state of the client.

Empathic improvisation

This is difficult to illustrate in a book or on a CD. It involves a therapeutic method that was first applied by Juliette Alvin where, typically at the beginning of a session, she would play (on her cello) an improvisation that empathically complemented the client's 'way of being'. In practice this means taking into account the client's body posture, facial expression, attitude on this particular day and previous knowledge of their personality and characteristics, and playing something to them that reflects a musical interpretation of their own way of being at that moment. It was intended by Alvin as a very empathic technique, not attempting in any way to change the client's feelings or behaviour, but simply to play them to the client without any hidden manipulation of their feelings. If a client comes into the therapy room agitated and upset, this mood can easily be incorporated into an empathic improvisation; the therapist is not trying to ameliorate or reduce the degree of distress which the client is currently experiencing but merely to play it back to them as a supportive and empathic confirmation.

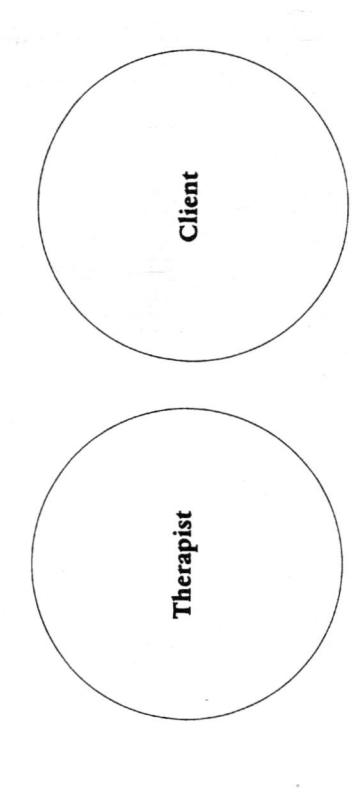

Figure 4.6 Two separate circles, representing separate musical identities, but with emotional empathy

Reflecting

This technique is well documented in Bruscia's 64 techniques and he defines it as 'Matching the moods, attitudes, or feelings exhibited by the client' (Bruscia 1987, p.540)

In reflecting, unlike mirroring, copying or matching, the therapist's music might be quite different from the client's as the purpose of this therapeutic technique is to understand and reflect back the client's mood at that moment, rather than be a more technical reflection of their music. However, there needs to be congruence in mood or emotional expression between the therapist's music and the client's music otherwise the method would cease to have any empathic effect.

Conceptually, we can see two separate identities of the participants in reflecting (client and therapist), in a relationship where they are separated musically, yet still congruent emotionally. Figure 4.6 illustrates the separation of the therapist and client circles.

CD23 demonstrates a client playing in a random, rather directionless rhythm on percussion instruments (drum and cymbal). Note that the therapist allows a short time to pass before beginning to reflect musically and empathically. This is an important part of the process:

Listen to the client's music before giving a response.

I frequently find myself reminding students in training and therapists under supervision that reflecting on your experience of the client's music is essential to be sensitive in response. There are sometimes patterns or characteristics that can help both in deciding the therapeutic method of response and the musical 'style'. The response the therapist gives in CD23 reflects the aimless and random style of the client's playing, using melody and harmony.

CD Example 23: Reflecting example 1 – therapist on piano, client on drums and cymbal

In the next example, the client presents a very different picture while playing the piano. Feelings of anger and frustration are present in the sharp, bunched chords the client is playing. There is an underlying sense of pulse, with accents and sudden changes in dynamics to reinforce the apparent irritation of the client. The therapist reflects these feelings with a melodic line on the xylophone.

CD Example 24: Reflecting example 2 – therapist on xylophone, client on piano

Two exercises are now presented on the CD, with the client playing piano in the first and temple blocks in the second. While these examples do not allow the reader to understand the actual emotional state or feelings exhibited by the client, they can be used by imagining what they could be, based on the music that is presented and trying to find a way to frame a response that is an empathic reflection of the music.

Exercise: Using CD25 and CD26, listen to each example for a few seconds, establishing in your mind the possible emotional state or mood of the music ('client'), and then allow your own emotional state to be affected by the music you are listening to. When you have become sensitive to the mood or emotion present in the music you are listening to, and your own emotional reaction to it, begin to play that emotional reaction on another instrument, reflecting the feelings that are present in the music, and present in yourself.

CD Example 25: Reflecting exercise 1 – client on piano

CD Example 26: Reflecting exercise 2 – client on temple blocks

4.4 Grounding, holding and containing

Grounding, holding and containing are all therapeutic methods that are extremely useful when applied with clients who have a very random or floating way of playing, and way of being. It is helpful where the client appears or sounds unconnected to their music, or the music lacks any stability, direction or intentionality. I have defined the process of grounding as:

Grounding: Creating a stable, containing music that can act as an 'anchor' to the client's music.

Examples of specific musical techniques that can be used in grounding include:

- strong octaves or fifths in the bass of the piano;
- steady pulsed beats on a bass drum;
- strong chords of a stable tonal nature using typically dominant and tonic chords;
- a simple ostinato.

Rhythmic grounding

Rhythmic grounding is a very useful way of providing a foundation to something the client is doing. Bruscia defines it as 'Keeping a basic beat or providing a rhythmic foundation for the client's own improvising' (Bruscia 1987).

An important aspect of rhythmic grounding is that it is not necessary to impose a meter on the client's rhythmic musical production. In fact, it can be quite constraining and directive to take the client's musical production and establish a specific meter such as 4/4 or 3/4 for what they are doing. Music can be pulsed but meterless, and quite often becomes more dynamic by the variable use of accentuation within a stable pulse. Another important aspect is to intervene with a stable and secure melodic or rhythmic pattern, quite often limiting your playing where a client's playing is rather full and complex. The process of limiting in the therapist's music is to provide a stable and understandable ground, and avoid adding to the potentially chaotic complexity of the client's improvisation (Figure 4.7).

CD27 is an example of a client playing randomly on the xylophone. The therapist then joins in on a drum and establishes a rhythmic ground to the client's music. You will hear the client begin to 'entrain' to the therapist's rhythmic ground and stabilize his or her own music accordingly.

CD Example 27: Example of rhythmic grounding – client on xylophone, therapist on a drum

Exercise: The next example on the CD (CD28) is a person playing a xylophone. Try to listen for any rhythmic patterns in the person's music, and then introduce a rhythmic ground. Remember, the faster or more complex the client's music, the more stable and limited must be the musical ground of the therapist. As this is an exercise requiring you to play with a CD example, potential for the person playing on the CD to 'adapt' to the therapist's grounding is clearly not expected. However it is a good exercise to practise finding ways of developing *matching into grounding*

CD Example 28: Rhythmic grounding exercise client on xylophone

Tonal grounding

Tonal grounding is a process where one establishes a tonal bass which acts as a foundation or 'anchor' to the client's music if it is predominantly melodic or harmonic and is wandering around. I define this as:

> *Tonal grounding: Providing an octave, fifth or harmonic chord in the bass that is congruent with, and tonally grounding for, the client's music.*

Bruscia defines this as tonal centring – 'providing a tonal centre, scale, or harmonic ground' (Bruscia 1987, p.535).

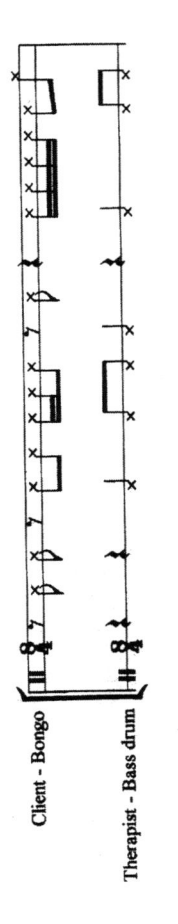

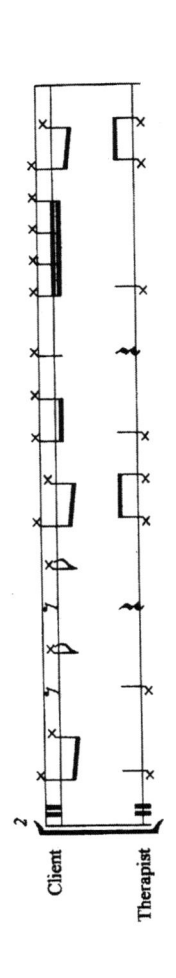

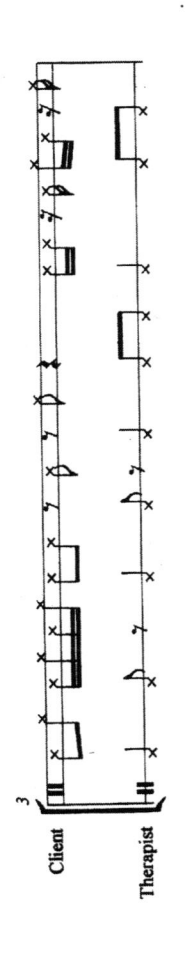

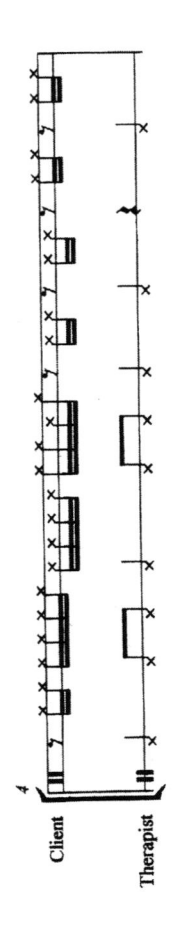

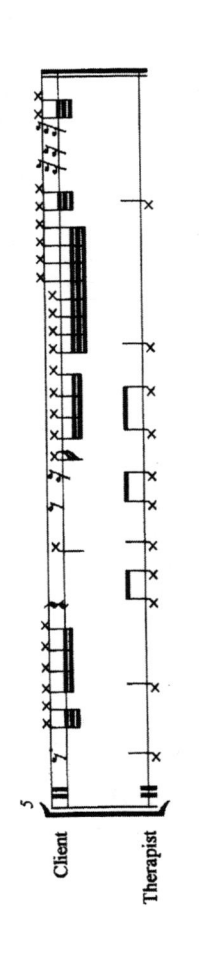

Fig 4.7: Rhythmic grounding – client and therapist on bongos and bass drum

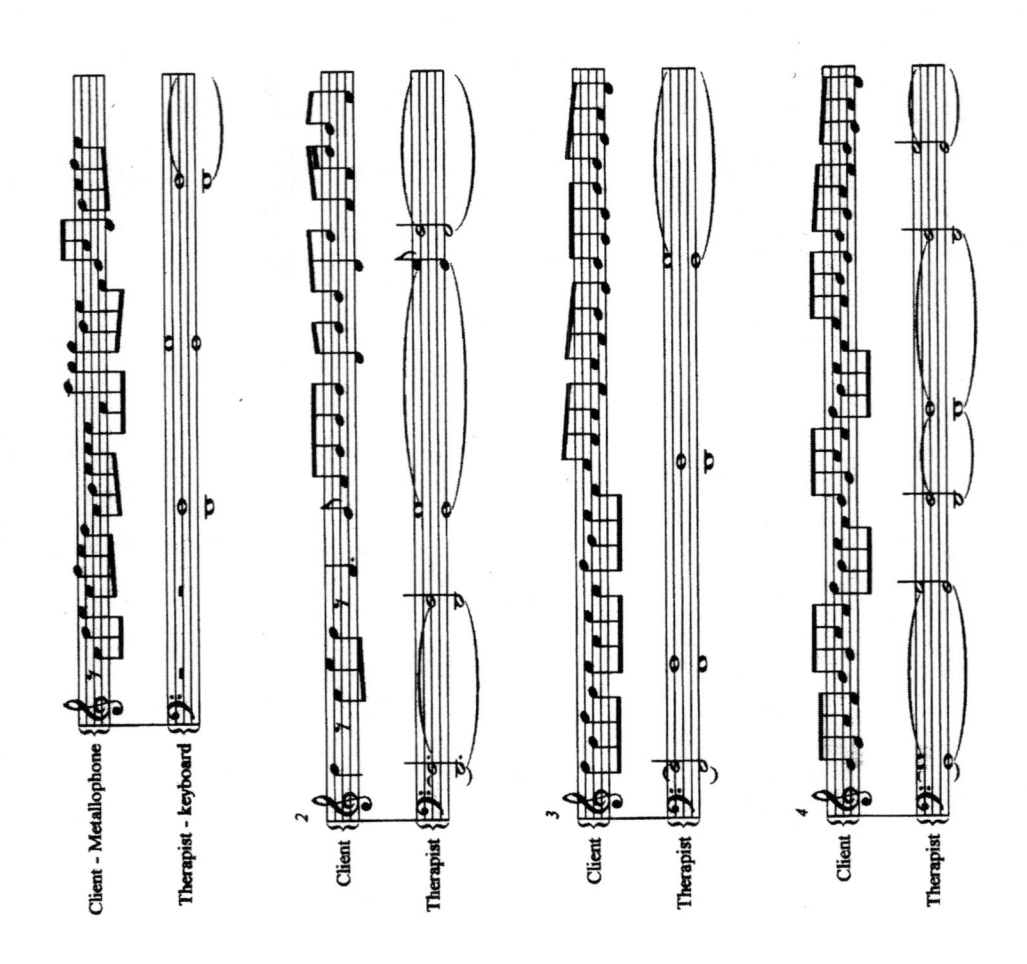

The musical example (Figure 4.8) illustrates this; a client plays a rather random, directionless melody on a metallophone which develops into repetitive patterns of falling thirds. The therapist intervenes with a tonal ground on the piano.

CD Example 29: Example of tonal grounding – client on glockenspiel, therapist on piano

Exercise: CD30 provides an exercise where a person plays music on a piano and as a duet partner you can work in the bass to provide some tonal centre for this. The technique involves analysing the type of music the person is playing and seeing if it falls within a key, or if a ground tone could be used as a tonal centre. For example, if the client is playing mainly the white notes of the piano, A minor, D minor and C major could be used as keys to provide a tonal centre. If the client is playing on the black notes, E flat minor and F sharp major can be used as the keys to provide the tonal centre (pentatonic).

CD Example 30: Tonal grounding – moving from diatonic to pentatonic

Harmonic grounding

Tonal grounding can be extended to *harmonic grounding*. This tends to involve either tonal harmonies (as in the two-chord improvisation) or pentatonic harmonies. As an extension to the use of fifths and octaves for tonal grounding, try using the CD30 exercise to engage with harmonic grounding.

Combined tonal and rhythmic grounding

Rhythmic grounding and *tonal grounding* can be combined to establish an even more secure musical foundation for a client. A good example of this would be to use a drone bass accompaniment figure to provide such a combined grounding foundation (Figure 4.9). The style could be given a 6/8 Celtic flavour by some suggestions from the therapist in the accompaniment, and then the harmonic ground can be enhanced with chordal structures (CD31).

CD 31 shows how the therapist maintains stability in the piano.

CD Example 31: Combined rhythmic and tonal grounding – client on piano, therapist also on piano

Holding and containing

Holding and containing are quite similar therapeutic methods. Basically, I employ holding as a therapeutic method and process where one provides a musical anchor to

Figure 4.8: Example of tonal grounding – client on metallophone and therapist on piano

a client who is ungrounded in his or her playing and whose music is random and without direction. Consequently techniques such as tonal grounding/tonal centring are going to be helpful in order to provide that anchor. It works well to use simple harmonic accompaniments as a holding 'tool' where the use of sustained sounds without attempts at interactive or dynamic music making provides the containing frame. The therapist's music would typically be slow, sustained and very stable. However, at the same time it doesn't have to force a pulse or a meter on the client for it to be good enough music for holding. Therefore I define holding as:

Holding: Providing a musical 'anchor' and container for the client's music making, using rhythmic or tonal grounding techniques.

Bruscia offers a different definition of holding, one that is more expanded to include the wider concept of the 'musical background', and also includes the concept that the technique contains the feelings of the client: 'as the client improvises, providing a musical background that resonates the client's feelings while containing them' (Bruscia 1987, p.536)

Containing implies a different process where the client's music is quite chaotic and may also be quite loud. Therapeutically, the client needs to be allowed to be chaotic, noisy, exaggerated (a good example would be an out-of-control child having a 'musical/emotional' tantrum). The therapist provides a musical container for the client's music, playing strongly and confidently enough to be heard by the client. One musical idea that can work well in therapy is to play at opposite ends of the piano with strong, stable octaves (CD32). Many other types of music could act as a container for the client's music, but it needs to be structured music that provides a pattern.

CD Example 32: Containing: Chaotic music contained by the therapist – client on cymbals, drums and xylophone, therapist on piano

4.5 Dialoguing

Music is a marvellous medium for engaging in different types of conversation or dialogue between two or more people. It is even possible, of course, to have a dialogue with oneself musically! I have not found a definition for *Dialoguing* in its application in music therapy as either a musical technique or a therapeutic method, although there are terms that describe some of the processes involved in making or developing a dialogue. I define dialoguing in the following way:

Dialoguing: A process where therapist and client/clients communicate through their musical play.

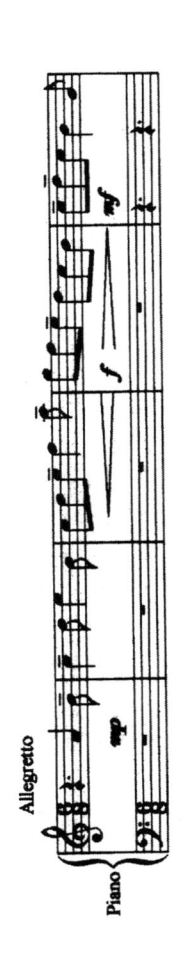

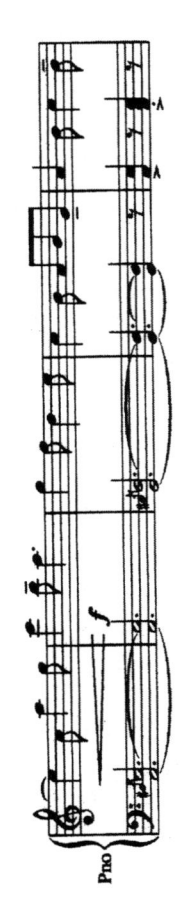

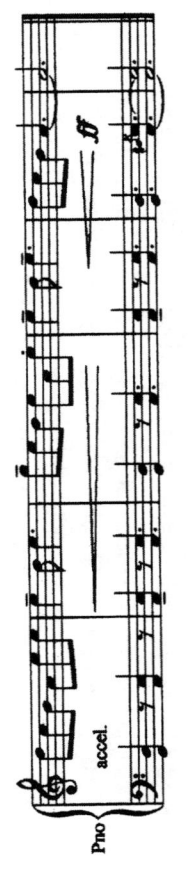

Figure 4.9: Combined rhythmic and tonal grounding – client on piano and therapist on piano ground, even when there are mismatches in the harmony between the client's 'jumping around' melody and the drone ground

There are two main forms of dialogue, which I define as follows:

Turn-taking dialogues: Making music together where the therapist or client(s) in a variety of ways, musical or gestural, can cue each other to take turns. This 'turn-taking' style of dialogue requires one or other to pause in their playing and give space to each other.

Continuous 'free-floating' dialogues: Making music in a continuous musical dialoguic exchange – a free-floating dialogue. Here participants (therapist(s) and client(s)) play more or less continuously and simultaneously. In their playing musical ideas and dynamics are heard and responded to, but without pause in the musical process.

To liken a dialogue to a conversation is probably the nearest and most understandable way of describing this process. Consequently, one can imagine that just as in a conversation, there are a number of ways in which the dialogue can progress:

1. Therapist and client(s) take turns to play, taking over immediately from each other.
2. Therapist and client(s) take turns to play with pauses in between 'statements'.
3. Therapist or client(s) interrupt each other.
4. Therapist and client(s) 'play at the same time' (talk at the same time) as each other.
5. Client(s) make(s) long statements; therapist gives 'grunt' or 'ah-ha' responses of very short phrases.
6. The therapist's musical style in the dialogue is very empathic (similar) to the style of the client(s) (or vice versa).
7. The therapist's playing in the dialogue is very oppositional/confrontational to the client(s) (or vice versa).

Ways to promote dialogue

Musical dialogues don't necessarily occur automatically or naturally in improvisational music making. In fact, some clients find it extremely difficult to engage in dialogues, either because they can't follow or respond to normal turn-taking exchanges (typical in autistic clients), or because they talk so much that they don't stop for long enough to listen to what somebody else has got to say (this can be typical in clients with Asperger's syndrome).

Before explaining more specific techniques for promoting dialogue, there are two clearly defined therapeutic techniques proposed by Bruscia that can be utilized:

Interjecting – waiting for a space in the client's music and filling in the gap.

Making spaces – leaving spaces within one's own improvising for the client to interject his/her own materials (Bruscia 1987, p.535).

Using these two methods naturally leads one into dialoguing and initiates the 'conversation' or 'argument' style of improvisational music making, where the playing together becomes directly communicative. Many clients may not understand or pick up the signals that help nurture dialoguing, and this can be helped through modelling. Modelling is a method that can be applied to many of the previously described musical and therapeutic techniques, and many of those yet to be discussed. Bruscia's definition of modelling is:

Modelling – presenting or demonstrating something for the client to imitate (1987, p.535).

This provides us with a quite specific (and clearly directive) method which is most useful where that type of direction is needed. I would like to suggest an extended and broader definition here in order to explain that something more than purely imitating occurs:

Modelling: Playing and demonstrating something in a way that encourages the client to imitate, match or extend some musical ideas.

In the music making that goes on in music therapy there are subtle or obvious ways of promoting the initiation, development and progression of a dialogue. These involve either musical cues or gestural cues.

Musical cues

- Harmonic cues: indicating that you are coming to the end of some musical 'statement' by playing either a perfect or plagal cadence (or even an interrupted cadence). The harmonic modulation in a musical statement can also sound like a question.
- Rhythmic cues: playing a rhythmic pattern that closes, following which it is obvious that there is a space or playing a rhythmic pattern that is symmetrical and therefore gives a clear indication of closure (also allowing space for a client to play next).
- Melodic cues: playing in ascending phrase, a phrase that indicates the end of a pattern, etc.

- Dynamic and timbre cues: there are many types of dynamic cues that could indicate a space for developing a dialogue. Accents help to establish a punctuation point; making a crescendo on a phrase to a climax indicates a point of stopping; making an accelerando to a point of stopping also indicates a pause which allows a space for somebody to say something; staccato playing following some legato playing may also indicate something coming to a conclusion.

Gestural cues

Given that musical cues can be rather subtle and are not necessarily attended to, especially by clients who enjoy making a lot of noise and playing continuously, it may be necessary to model the dialogue idea through giving a gesture. The idea is to indicate a space where you would like the client to start playing (or continue playing) on their own in order to develop the dialogue. Therefore you can introduce some of the following ideas:

- Show you have stopped playing in some way, by taking your hands from the instrument or 'freezing' at the instrument so that you are not moving at all and looking as if you are waiting for the client to stop before you can play again (very effective with children when they catch on to the idea as it gives them a strong sense of being 'in control'!)
- Turn to look at the client and take your hands off the instrument.
- Use eye referencing to indicate that you are going to play and then eye reference the instrument to encourage the client to play.
- Point and indicate whose turn it is to play.
- Use physical prompts, either to encourage somebody to start playing, or to encourage them to stop playing:

Starting to play:

- nudging behind the elbow;
- supporting under the elbow;
- supporting under the wrist;
- taking a hand and helping a client to play.

(This is a graduated list of responses from a very gentle prompt to a hand-over-hand modelling.)

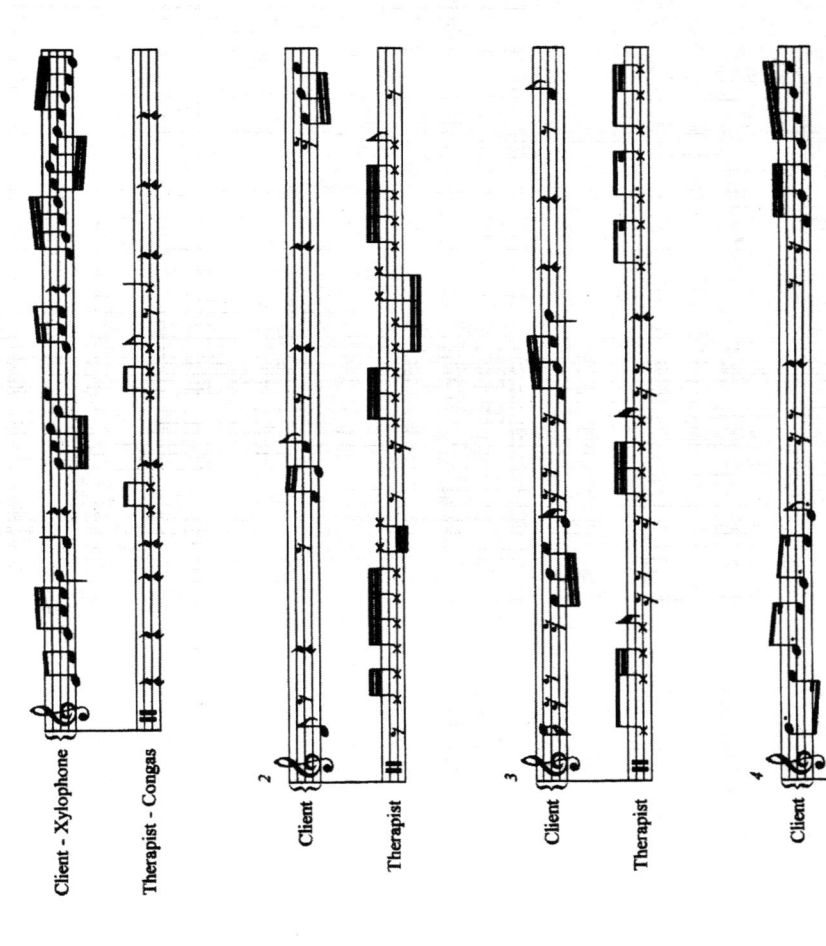

Figure 4.10: Example of Dialoguing – client on xylophone, therapist on congas

Figure 4.11: Example of conversational dialogue using variable phrasing, continued on next page

Stopping playing

- putting the hand out in a stop position;
- reaching over and almost touching the hand of the client;
- reaching over and holding the beater or instrument that the client is using to play for a short time;
- reaching over and stopping the client playing physically by holding their hand; taking an instrument away while you interject a short phrase and then handing the instrument back.

(This is a graduated list ranging from gestural cues to physical direction.)

Figure 4.10 illustrates an emerging dialogue beginning with a client playing on a xylophone, without pulse, and shows how the therapist gently interjects, makes spaces for the client, then uses rhythmic patterns to develop the dialogue.

CD Example 33: Dialoguing 1 – client on metallophone, therapist on xylophone

The techniques described above range from subtly to strongly directive. Direction in some form is sometimes necessary in order to build up, through modelling, the process of musical dialoguing or turn-taking. I am often asked how one can develop communicative musical dialogue with clients who have perseverative and repetitive playing, who seem to be unable or unwilling to leave any space in their musical production to allow a dialogue to develop. The ideas described above are typical in the techniques I have found helpful to model, initiate and develop dialogue. However, one also needs to take into consideration the instrument chosen and the physical playing style. Clients who play repetitive pulses on drums may do so because the motor movement (also described as sensory motoric playing) is what they are interested in doing, and there is little or no musical or communicative intentionality. All the above techniques may prove futile in the face of such playing, and changing instruments may be the best way to break down obsessive patterns of playing and introduce dialogue.

Phrasing, interrupting, pausing and talking at the same time

Having begun to develop dialogue, the patterns that emerge can sound more and more like a conversation when attention is paid to phrasing, interrupting, pausing and talking at the same time. Phrase lengths vary – especially where one person is doing most of the talking, and the other is merely acknowledging or confirming with an 'uh-huh' response. So, in musical dialogue, these patterns of conversation can increasingly represent the prosody and phrasing of speech, with accents, inflection, interruptions and sometimes even talking at the same time. In the process of dialoguing – whether through a rhythmic or a melodic exchange – the potentials of varied phrasing will add significantly to the communicative character of the dialogue.

Figure 4.11 illustrates this, and CD Example 34 shows how all the dynamic aspects of interpersonal communication can be present in a musical dialogue. Given that music therapy is a medium through which 'communication' takes place through musical exchange, dialoguing is a very important and valuable technique to support and engage a client.

In the real world, communication and dialogue between people can frequently turn into a heated debate, perhaps even an argument. Polite turn-taking gives way to interrupting, increasing accents, 'rude' sounds, shouting, losing tempers – everything a good healthy argument should have! CD34 illustrates the musical dynamic of dialogue that becomes an argument, and as music therapy allows people to say something in music (in an argument) that would be unacceptable in words, this is a valuable tool in therapy work to draw out emotional attitude and affect.

CD Example 34: Dialoguing 2: Conversations and arguments! Therapist on xylophone, client on African split drum and djembe

Continuous 'free floating' dialogues

When working with clients who play quite continuously, repetitively, perhaps even obsessionally, and have difficulty in stopping to listen, the therapist's option is to try to promote or engage with the second type of dialogue method described above – the continuous 'free-floating' dialogue. Here, the therapist can listen to and echo musical ideas, themes, motifs and dynamic patterns of the client, attempting to build up a dialogue of musical ideas within an ongoing improvisation.

It cannot be compared with a conventional conversation, where turn-taking is a typical element. In the free-floating dialogue, the musical genre of opera is represented, where two (or more) people can be simultaneously contributing to an exchange, sometimes singing about two different things at the same time, yet with a necessary musical connection through melody or harmony. It happens frequently in improvisations, and this kind of instantaneous reciprocity and shared understanding builds up between client(s) and therapist, and acts as a framework for communicative experiences. The subtlety of this type of interaction is such that it is not always possible to be aware of how it is happening while it is going on, and only with later audio or video analysis can one recognize the presence of a subtle and developing dialogue. CD35 gives an example of just such a dialogue, where the therapist uses

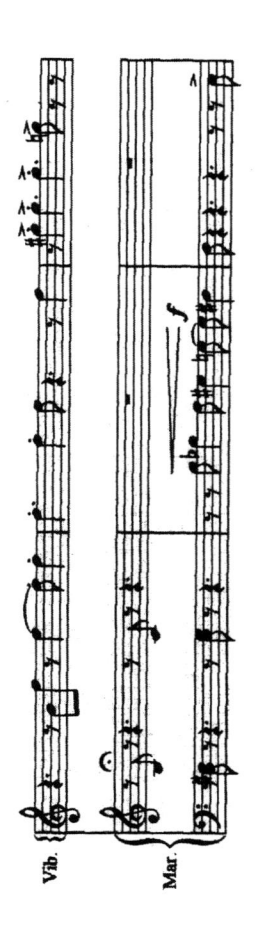

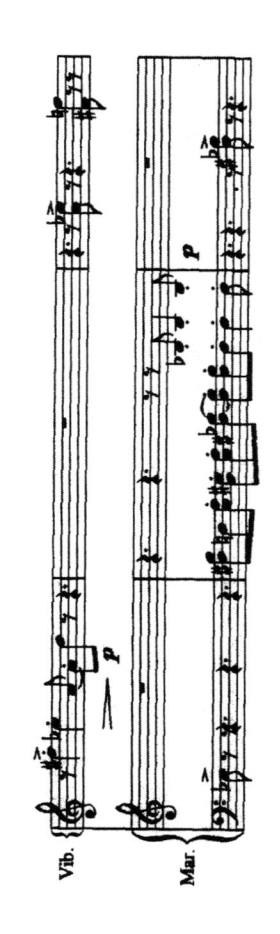

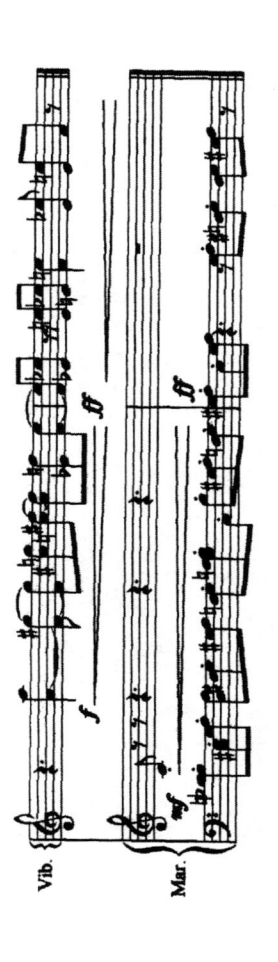

Figure 4.11: Example of conversational dialogue using variable phrasing, continued

the xylophone to match, and then dialogues with a client's continuous playing on a drum.

CD Example 35: Dialoguing 3: continuous 'free-floating' dialogue – therapist on piano, client on xylophone

4.6 Accompanying

Accompanying is one of the most useful and important of the supportive techniques in improvisational music therapy. I often recommend its use when one has established a framework for clients to use or where a client is particularly autonomous and wants to take a soloist's role in the music making.

I define the therapeutic method of accompanying as:

Accompanying: Providing a rhythmic, harmonic or melodic accompaniment to the client's music that lies dynamically underneath the client's music, giving them a role as a soloist (Wigram 2000b).

Accompanying is a frequently used method for joining in with a client's music where the message one is giving is of support and empathy. The definition refers specifically to the idea that the music lies 'dynamically underneath', and this typifies the quality of 'accompanying' and gives it strength as a supportive music. If the client plays *f* then the accompaniment is going to be *mf*. If the client plays above middle C in the tonal range, the accompaniment can be placed lower, although it is possible to work with a bass lead and an accompaniment in the higher register.

Accompaniment style music, certainly on the piano, needs to have certain characteristics:

- to be simple and repetitious;
- to be a short rhythmic or harmonic sequence that is sustained;
- to continue in a stable way despite some changes in the client's music;
- to be sensitive to pauses or small developments in the client's music.

Typically, accompaniments can be (either tonal or atonal) um-cha-cha (3/4 waltz) style or um-cha-um-cha (2/4 and 4/4 common time) style. Figure 4.12 gives us an example of this type of accompaniment, in both a tonal and an atonal frame.

However, there are also some important other types of accompaniment. The 2-chord improvisation that was exemplified in Chapter 3 is a good sequence to use for an accompanying style, as is the Spanish 2–8-chord sequences that will be explained in Chapter 6 under frameworking techniques.

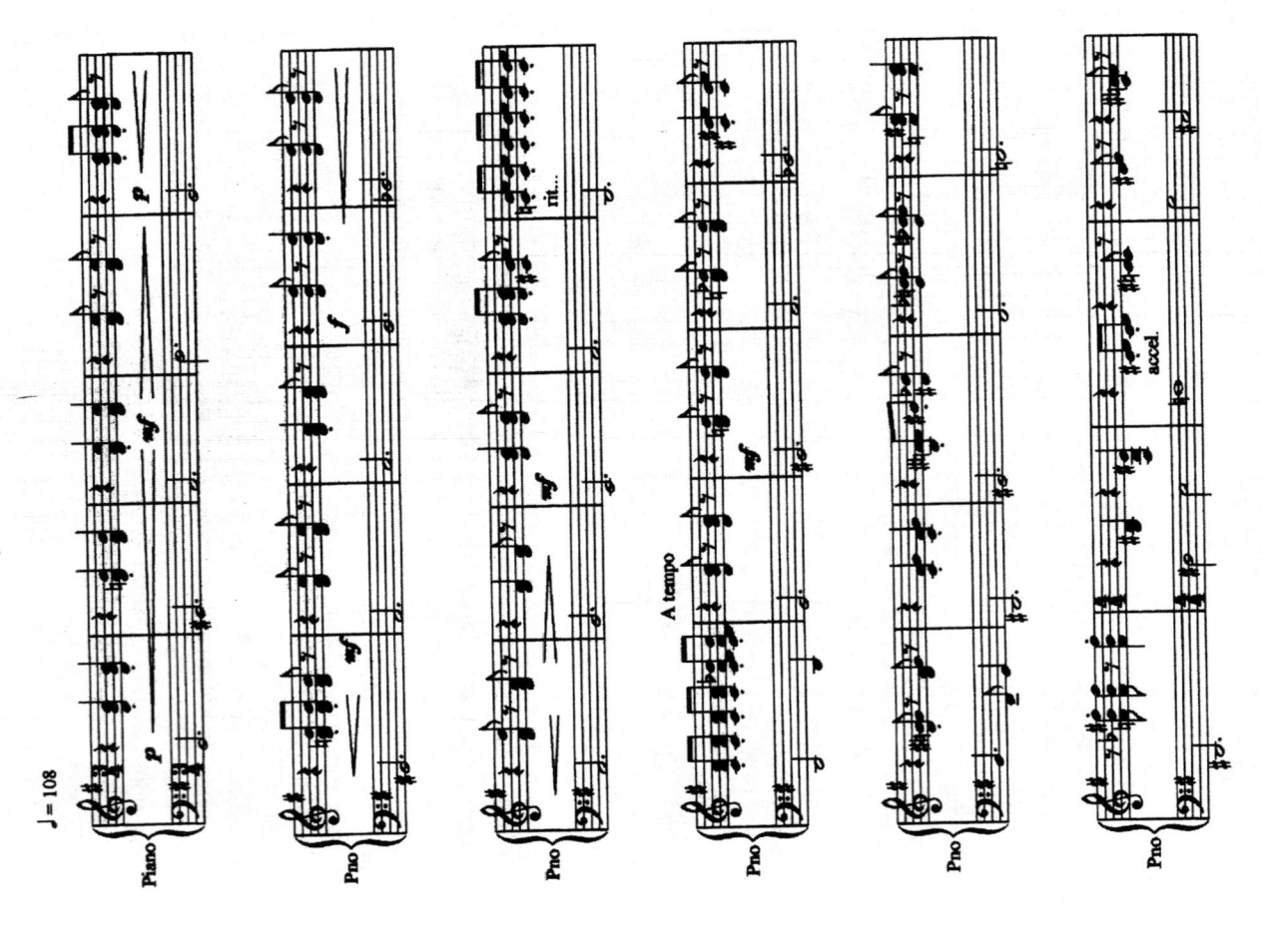

Figure 4.12: Example of 3/4 and 4/4 accompaniment style using tonal and atonal frame, continued on next page

BASIC THERAPEUTIC METHODS AND SKILLS

CD36 gives an illustration of a client starting to play randomly on a xylophone and glockenspiel while the therapist introduces an accompaniment style using (at first) two chords to support it, then developing some accompaniment effects.

CD Example 36: Accompanying – client on xylophone and glockenspiel, therapist on piano

Most of these accompaniment methods can be equally effective on guitar or other harmonic instruments (harmonica, accordion, autoharp, organ, synthesizer). Purely rhythmic accompaniments can also be generated, and are especially effective in providing a supportive frame. The most important characteristic of this therapeutic method to remember is your supportive role, allowing the client to take the lead, playing more softly, with stability and repetitious motifs of figures, and perhaps with a thinner, sparser texture.

> Exercises: Try making different types of accompaniments to the following styles of playing using the examples on the CD with which to work:
>
> **CD Example 37**: Accompaniment exercise 1 – a wandering treble melody by a client on a piano where they play first of all only on the white keys and secondly only on the black keys
>
> **CD Example 38**: Accompaniment exercise 2 – a client playing an accented, rhythmic and pulsed melody on xylophone and metallophone, that breaks out of meter halfway through
>
> **CD Example 39**: Accompaniment exercise 3 – a client playing some rhythmic patterns on a drum
>
> In all three exercises try formulating accompaniments using different instruments such as piano, guitar or drums/percussion.

4.7 Summary and integration

These are some basic therapeutic methods that need to be practised in order to acquire both the technical and therapeutic skills to use them. As can be seen, they start to incorporate the musical techniques that are adapted to fit the intention of the method. The exercises suggested in Chapters 3 and 4 are designed to allow the reader a chance to practise these methods using either piano or other instruments. Many of these musical techniques and therapeutic methods will be revisited in later chapters because improvisation is not undertaken with clients through isolated methods, but through a sequence (sometimes fast-moving) of different methods and musical techniques.

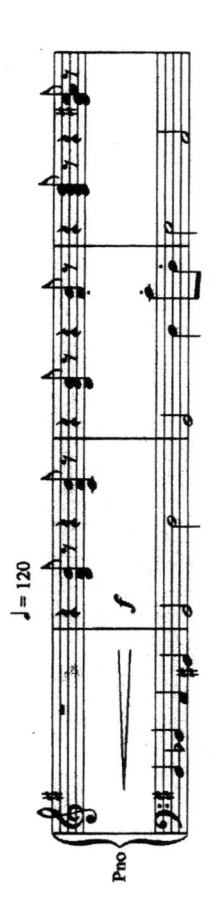

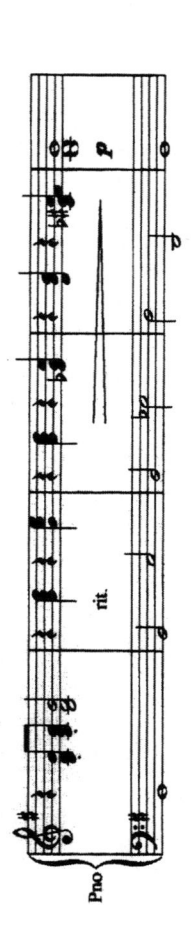

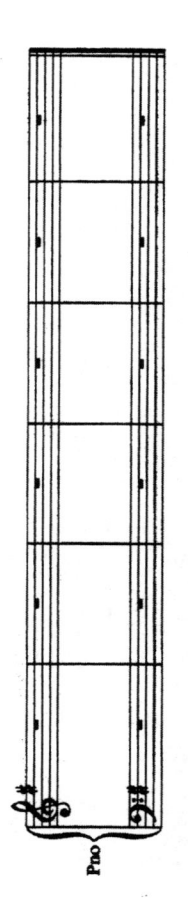

Figure 4.12: Example of 3/4 and 4/4 accompaniment style using tonal and atonal frame, continued

The last part of this chapter is therefore concerned with the integration and sequential process of linking together these methods to illustrate how one can move through a therapeutic sequence of events with a client. As has been stated earlier, *matching* is a logical and empathic place to start with a client. However, in therapy we don't approach our clients with some predetermined plan of intervention, at least not in improvisational music therapy. The spontaneous experience, adapting and responding on a moment-by-moment basis to the interactive process, requires us to maintain a free-flowing flexibility in the application of therapeutic method.

The last example, illustrated only as audio example CD40, shows how one might move through three or more methods in an improvisational interaction with a client.

Matching ⟶ Accompanying ⟶ Dialoguing ⟶ Containing ⟶ Matching

The client is playing a xylophone, and begins with rhythmic, melodic fragments. The therapist matches, and the engagement begins. As the client grows more confident, the therapist takes the role of accompanist. A little further on, the therapist takes an initiative by making spaces and interjecting, and introduces the idea of dialoguing. The client works with this, but as the dialogue builds up dynamically to an argument, the therapist adapts to a containing approach. As the client's music loses some of its intensity and energy, the therapist follows and returns to a final empathic section of matching.

CD Example 40: Example of integrating therapeutic method and musical technique

So far, the techniques and methods recommended for both practising and developing within an improvisational model for use in clinical work have concentrated on identifying specific techniques using musical parameters and therapeutic method. Most of the examples and the exercises recommended have involved a form of improvisation where the music is spontaneously created, using some simple play rules.

Frequently, when working with musical material, one wants to develop a style of improvisation that fits something that the client may be doing or to create a particular type of musical frame for some specific purpose. I call this method of work 'frameworking'. In addition, we are constantly faced with the need to find ways of making changes in the music, making a transition from playing in one way to playing in another way. The development of these transitions is a critical part of music therapy skills (and in fact is used very widely by musicians, composers and others to connect together different types of music).

In order to move to the next stage of the process of developing improvisation skills, I will describe and give examples of both frameworking and transitions and then explain a number of exercises that can be used to develop these methods.

6

Songwriting to Explore Identity Change and Sense of Self-concept Following Traumatic Brain Injury

Felicity Baker, Jeanette Kennelly and Jeanette Tamplin

Rehabilitating physical, cognitive and communication functioning following Traumatic Brain Injury (TBI) is an intensive, exhausting and highly emotional task for children, adolescents and adults. Successful rehabilitation relies on clients maintaining high levels of motivation. This is often difficult to achieve, however, when emotional responses to the trauma have an adverse impact on the client's levels of motivation. Over the past 12 years, the authors have successfully employed songwriting interventions with children, adolescents and adult TBI clients to facilitate the adjustment process and to help maintain their motivation for therapy. This chapter will outline the approaches we used in conducting therapeutic songwriting with TBI clients. We first highlight the specific adjustment issues faced by TBI clients to provide a sense of the complexity of dealing with these clients' adjustment processes. Included in this chapter is an explanation of some of the cognitive impairments typically acquired by TBI clients and, more importantly, their impact on the adjustment and songwriting process. Following this, our protocol for writing songs with clients is outlined and this is illustrated through two case examples – one with a late adolescent client and the other with a paediatric client.

Confronting and adjusting to change

Traumatic brain injury is often the outcome of a sudden and unexpected event which results in damage to the brain. Given the unexpected hospitalization and threats to independence, it is not surprising that clients experience emotional crises and often undergo a lengthy adjustment phase. Clients have to adjust and cope with significant life changes that involve accepting many losses: loss of independence and functioning, loss of control, loss of former body, loss of financial status, loss of many roles, loss of future hopes and dreams, and loss of ability to participate in preferred leisure activities.

Several theories have emerged about how and when adjustment occurs and what variables influence the process. Wright (1960) viewed adjustment to disability in terms of it reflecting the interaction between a person's value system, level of emotional maturity and acceptance of self, and mental health status.

Olney and Kim (2001) suggest adjustment is a staged process which includes:

- the response to the initial impact
- defence mobilization
- the initial realization
- a period of retaliation
- reintegration and adjustment which is characterized by confidence, contentment and satisfaction.

Adjustment involves the formation of an identity that integrates all aspects of the self, as well as an understanding at multiple levels of the meanings and implications disability has on the person's life. Major themes arising from such processes include: how individuals describe their difficulties; how they cope with specific limitations; and how they manage their identity and integrate their identity as a person with a disability into a cogent sense of self.

With specific reference to TBI, Simpson, Simons and McFadyen (2002) propose that the major challenge faced by people after TBI is reaching an understanding of exactly how the injury has affected their cognitive and psychosocial abilities. They experience an uncertainty about the full impact of the TBI throughout the period of recovery, rehabilitation and longer-term adjustment. The full impact of the injury may remain 'hidden' for some time.

TBI clients may be long-term patients and it is not unusual for them to be cared for within the hospital ward for up to two years. There are two distinct periods within the recovery period when clients are most vulnerable and confronting adjustment issues:

Stage 1 The first period usually occurs as they approach the stage of rehabilitation where progress begins to slow and there is a growing realization that a full recovery is becoming less likely.

Stage 2 The second period of vulnerability occurs between 6 to 12 months after discharge. On initial discharge, there is excitement about leaving the hospital and returning home. However, as this excitement wears off and the reality of long-term life changes is contemplated, boredom and depression may ensue.

Appropriately timing the inclusion of this method into therapy programmes with this population is crucial, and it is at these particular two phases of recovery from TBI where songwriting interventions can be especially valuable. Clinicians need to consider the appropriate timing in which to encourage reflection and adjustment through songwriting. The music therapist has an ethical duty to safeguard the emotional well-being of clients by not raising issues of which the client is not yet aware or emotionally ready to deal with. When client recovery is active, full participation in rehabilitation is essential. During this important phase of treatment, music therapists also need to make a clinical decision as to whether exploring adjustment issues are appropriate at that time. Reflecting and reviewing one's situation can lead to temporary crises when moments of insight occur and this may be detrimental to a client's treatment programme. More appropriate is the inclusion of self-reflection through songwriting when the client's rehabilitation is being hampered by negative emotional responses. In this situation, a client needs to work through these issues in order to maintain motivation for continued therapy.

Exploration through song

Coping and adjusting to trauma has been promoted within our therapy programmes by facilitating client exploration of thoughts, feelings and reactions to their acquired injury through songwriting. In analysing the lyrics of 82 songs written by clients with TBI, several themes emerged within the songs which could be directly related to aspects of patient adjustment to injury. These have been detailed in a number of our recent publications (Baker, Kennelly and Tamplin in press[a], in press[b], in press). In particular, clients described feelings about, and responses towards, their present situation including the distress and pain involved in the hospitalization process and the feelings of isolation, dependency, helplessness and anger associated with their current situation. Many clients voiced concerns about how their physical and cognitive impairments caused others to view them, thus articulating confronting and painful issues related to body image (Charmaz 1995). Positive experiences were also included within songs, particularly experiences related to memories about, and reflections upon, significant others. When connecting with the positive aspects of the past and those people who were or had been playing an important role within their life, clients were able to grieve losses; a necessary step in the adjustment process. These losses may include loss of role within family, school, workplace and social circle; loss of cognitive, communication and physical function; loss of identity; and loss of financial security. The expression of past experiences also facilitates some relief from the overwhelming and often negative feelings related to their present situation. Grappling with the uncertainty of the future was also voiced within a number of the songs written by clients.

Communication impairments: inhibitors of adjustment

Language and communication impairments greatly inhibit the songwriting process and the therapeutic process. Clients with TBI may demonstrate a range of communication impairments including aphasia, dyspraxia and dysarthria:

- *Aphasia.* Neurological damage causing aphasia leads to impairments in word-finding and language. Clients with aphasia or dysphasia know what they want to say but can't find words or language necessary to communicate and express themselves accurately.
- *Dyspraxia and/or Dysarthria.* Neurological damage causing dyspraxia and/or dysarthria leads to impairments in articulating speech. Here, clients may be unable to clearly articulate the words chosen to express their thoughts and feelings. It may take lengthy periods of time to understand what the client is saying due to the need for repetition or the length of time required to articulate a message (due to initiation problems or fatigue).

When neurological damage results in an inability to verbally articulate or where speech is not the most efficient means of communication, alternative devices or strategies to communicate are used. Holding long, detailed discussions with clients about their situation may be beyond their abilities and the time frame of a session. Further, these clients often fatigue easily. Clinicians need to be aware that a client may not be accurately expressing the full range of their feelings due to these limitations.

Cognitive impairments: inhibitors of adjustment

Adjustment to disability involves understanding, exploring, responding and working through a range of loss and grief issues and forming a new identity. As outlined by Wright (1960), in a non-brain-injured person this staged process is dependent upon a person's value system and their level of emotional maturity. Therefore, the process of adjustment is inherently different for children than for adults. Further, the process is dependent upon various intellectual processes, implying that the greater the cognitive impairment, the more difficult the process of adjustment is. In particular, a client with a TBI may display a combination of cognitive impairments which directly affect and complicate the process of adjustment.

Poor insight is a common problem in clients with TBI. A client may be unaware of their impairments or the implications of their impairments. Consequently, adjusting to changed circumstances is challenging when a client is unable to perceive impairments. For example, a mother might continuously express a desire to go home and attend to the daily needs of her children. However, she has difficulty connecting her physical impairments, such as her inability to walk, with her inability to carry out specific tasks required in her parenting role.

Short-term memory problems impact on the therapeutic songwriting process. Clients may have difficulty resolving issues when unable to recall exploring these issues on previous occasions. They may raise the same issues at the beginning of each session without recalling sessions when the issue was previously raised. The therapist needs to be sensitive to the emotional intensity of the clients' self-expression as they may believe that they are talking about these experiences for the first time. Surprisingly, many clients with severe memory deficits can recall songs that they are writing in therapy from session to session. Often they will not recall the song if asked about it, but will spontaneously sing and recall the words of the song when it is sung. This reinforces the strong link in the brain between music and emotion and its value in helping clients recall experiences and participation in earlier sessions. This assists the client to move through the stages in the adjustment process.

The process is further complicated by gaps in long-term memory whereby clients may have difficulty recalling details they wish to include in the songs they are writing. For example, one client had a memory gap for the year prior to his accident. Within this time period, his wife had given birth to his baby and consequently he was unable to recall this important event in his life.

Additional cognitive problems found during the adjustment phase are:

- poor concentration
- poor attention
- limitations in planning and organization skills
- limitations in problem-solving, initiation, and abstract thinking
- inability to learn new information
- perseverative tendencies.

All these difficulties may impede the adjustment process. Some of these problem areas are further addressed in the next chapter, by Baker.

Introducing songwriting as an intervention

There are many different methods of approaching the songwriting process when focusing on emotional expression and adjustment issues for clients with a TBI. Some approaches are more appropriate than others for clients with particular needs and preferences. Often the clinician chooses the most appropriate way to introduce the songwriting process for a particular client, based on an assessment of cognitive functioning. It may be difficult for a client with cognitive difficulties to understand and remember a range of options for how to approach songwriting. To avoid confusing the client and to maintain the therapeutic aim of the songwriting intervention (i.e. assisting the adjustment process), it may be more appropriate for the clinician to be more directive in the process. The clinician might ask the client to express an idea and then immediately provide an example of how this idea may be represented as a song line. The client is provided with immediate feedback as to how the song will be created. The clinician may be able to use previously gained information about the client's musical preferences and use the intensity of emotion expressed by the client as a guide for how to represent the lyrics musically. For example, angry lyrics could be supported by loud, driving chords, or grief may be represented by a gentle accompaniment of slow arpeggiated chords. In this way, the client and clinician work together, 'unravelling' the song in the moment.

For clients with fewer cognitive difficulties, a range of options for how to begin the song may be presented and the client is able to choose the approach with which they feel most comfortable. Many clients don't feel confident with music composition and need to be supported and guided through the early stages of the song creation process. This may involve discussion of the potential therapeutic benefits of writing a song, including:

- identifying and externalizing emotions
- communicating to loved ones
- self-motivation and encouragement
- simply telling their story.

Songwriting can create an alternative way to approach reality and may precipitate a change in thinking. It also encourages and promotes growth and self-awareness (Glassman 1991).

In some cases it may be appropriate to play an example of a song that another patient has written (after consent has been given) to illustrate the process and result.

Many patients feel most comfortable starting with lyric creation. In this case, the clinician may ask the patient if they would prefer to use the music to a song which they already know or like and write new lyrics (song parody), or whether they would like to create a completely new song. It is often necessary to reassure a patient that the therapist is able to provide as much musical support and guidance as necessary in the composition of the music, should they choose to write an original song. Providing the opportunity for choice throughout the songwriting process is imperative as it allows the client to have as much creative control as possible. Empowerment is particularly important for patients with a TBI as much of their lives are beyond their control, which may lead to the development of an external locus of control (Fenton and Hughes 1989).

When to introduce songwriting into therapy

The songwriting process is rarely started in the first music therapy session when addressing emotional and adjustment issues for patients with a TBI is a main goal of therapy. In many cases, a patient who has been attending music therapy for some time may begin to explore these issues as more insight into their situation develops or as a discharge date draws closer and the reality of ongoing disability sets in. For a patient who is already involved in a music therapy programme, the move into songwriting can be a natural progression. The rapport developed over time between the client and therapist provides an environment of trust and support. In addition, the therapist may develop greater insight into a patient's personality and key issues over time, as result of working together during the rehabilitation process. For a patient who is newly referred to music therapy specifically to address emotional and adjustment issues, often the first session is used to develop rapport, build a sense of trust. tastes, allow the client to tell their story and build a sense of trust. To introduce songwriting in this first session may often be too confronting and scary for the new client; however, it may be appropriate to talk about the songwriting process, present different options for how to write a song and discuss possible song topics or themes. This then allows the client time before the next session to prepare mentally and emotionally to write a song. For some clients, however, it may be appropriate to begin the song in the first session, particularly in the case of frontal lobe damage and where memory problems and other cognitive issues are present. Similarly, in a situation where issues arise during the therapy session for a client, it may be appropriate to 'strike while the iron is hot', and capture ideas as they are presented, rather than delay the process until the next session. It can also be useful for the client to take away some record of a song's progress at the end of a session. This may be in the form of a written copy of the song ideas and/or lyrics or a recording of the unfinished song so that he or she can prepare new ideas or changes between sessions.

Beginning a new song: lyric creation

The most common way to start writing a new song is to brainstorm ideas and record these on paper. This may often follow a period of discussion where the client is encouraged to talk about issues that are important or troubling and express their feelings about these issues. The client is then encouraged to select a topic or theme that has arisen out of this discussion from which ideas for lyrics may be generated. This process, entitled Therapeutic Lyric Creation (TLC), often follows a fairly standard format and has been developed over time by the authors:

Stage 1 Generate a range of topics to write about.

Stage 2 Select a topic for further exploration.

Stage 3 Brainstorm ideas directly related to the chosen topic.

Stage 4 Identify the principal idea/thought/emotion/concept within the topic (which functions as the focus of the chorus).

Stage 5 Develop the ideas identified as central to the topic.

Stage 6 Group related points together.

Stage 7 Discard the irrelevant or the least important points.

Stage 8 Construct an outline of the main themes within the song.

Stage 9 Construct the lyrics for the song.

In building client confidence with songwriting, it is often most appropriate to start with this general brainstorming of thoughts and ideas. This allows the flow of ideas to begin and expression is not impeded by the need for lyrical structure yet. In spite of a client's ability (or non-ability) to write, it is often best if the therapist scribes the ideas as this allows the client to talk freely and without interruption to the thought process. The therapist should take care to transcribe what the client says verbatim, so as to preserve the integrity and authenticity of the client's ideas. These ideas may be reworded or reorganized later on in the process to fit into a lyrical structure. The therapist should also provide the appropriate degree of support to facilitate the client's expression of ideas. This may consist of asking questions about statements that the client has made or asking the client to expand on certain statements. It is also therapeutically important for the therapist to validate and encourage clients in their expression of significant personal issues. For clients who have difficulty with initiation, this initial process may be more productive if the therapist takes the role of an interviewer and asks the client key questions about issues which have been highlighted as being significant. For example, a client may make a statement such as 'I hate being in hospital', but have difficulty explicating why he or she hates it, or providing more information about his or her emotional responses. In this situation, the therapist may ask open-ended questions such as 'what is it, in particular, about being in hospital that you dislike?' or 'describe to me what it's like for you to be here'.

Fill-in-the-Blank and Song Parody Techniques

Alternatively, a Fill-in-the-Blank Technique (FBT) using a familiar song may be adopted. This technique has been previously described in the literature (Freed 1987; Goldstein 1990; Robb 1996). A song that the client relates to may be used and adapted to make it more personally relevant. For example, a song by Moving Pictures could be adapted and presented as follows:

What about me? It isn't fair
I've had enough
Now get me out of here
Can't you see? I want to be free
But every day I'm stuck in here.

Here, clients complete the lyrics by including words or phrases that were brainstormed earlier in the process. This technique provides more structure for clients who may have difficulty expanding and organizing simple ideas. It can also provide direction for the lyrics and may serve as a beginning point for a client who is having trouble getting started.

Song Parody Technique (SPT) uses the music of a pre-composed song whereby the lyrics of the original song are completely replaced by client-generated lyrics. In many cases, a combination of these two techniques is employed.

In our clinical experience, SPT and FBT are the most commonly chosen and adopted methods with paediatric patients. Due to their developmental stage, many paediatric patients have not developed an individual music identity separate from their peers. Therefore, they are drawn to specific popular songs and musical artists. The very nature of this technique avoids the need for paediatric patients to make decisions about musical elements which may be too abstract for them at this developmental stage. This may be further compounded by impaired cognitive functioning as a result of a TBI.

Song Collage Technique

Another technique, which can be helpful for clients who have difficulty identifying or articulating their emotions, is the use of Song Collage Technique (SCT). This technique involves the client looking through music books or the lyric sheet within CD covers and selecting words or phrases from pre-composed songs that stand out, or have personal significance to them. The therapist facilitates this process by presenting a selection of songs which he/she considers contain meanings or descriptions of situations with which the client may resonate or identify. In these situations, identifying with the messages of other songs can enhance this therapeutic process.

The clients can then add ideas to these words and phrases and reorganize or reword them into their own song lyrics.

The collection of words and phrases is similar to the brainstorming process mentioned earlier. In a similar way, like ideas are then grouped together and reordered to suit the client's preferences. The therapist encourages and supports the client in changing any necessary words, expanding ideas and adding phrases to link the different points within the song.

Use of Rhyme Technique

The use of rhyming lyric patterns herein termed the Use of Rhyme Technique (URT) can be employed to create structure in a song. The therapist should make a decision whether or not to introduce this option based on the client's cognitive abilities. If a client is able to generate lists of words that rhyme with key words which they have included, then this technique is a good way to expand and organize song ideas, for example sad/bad/mad/glad/dad.

If a client writes a song line such as 'being here just makes me sad', a list of words rhyming with sad (bad, mad, glad, dad, had) can be generated and one of these words that fits in with a previously brainstormed idea can be used, or the rhyming word can be used to generate a new idea. For example, 'being here just makes me sad, but I'll take the good with the bad', or 'being here just makes me sad, but I remember the good times that I've had'. Depending on the clients' cognitive abilities, the therapist may suggest phrases using the rhyming words that the clients have generated or the therapist may generate a list of rhyming words and ask the clients which word they relate to most and ask them to create the next phrase ending with this word.

Music composition

Often the music is created after at least some of the lyric creation has been completed. This is generally because many clients attending music therapy are not musicians themselves and feel most comfortable with the lyric creation part of the songwriting process. It is important to build the clients' confidence with a task with which they are more comfortable before introducing a more challenging or demanding task. The specialized skills of the music therapist are employed in explaining the music creation process and involving the clients in this process to the greatest degree possible. At the start of the songwriting process, clients are given the option of writing their own music, or using the music to a familiar song of their choice to which to write their own lyrics (SPT).

SPT is useful when clients feel daunted by the idea of writing music for their song, or when clients have difficulty conceptualizing how to structure a lyrical line without music. The use of pre-composed music can provide this sense of structure and gives the client guidelines for how many words or syllables for a line and how many lines for a verse or chorus. For some clients this sense of a familiar structure provides a feeling of security; however, for others it may be too limiting. It can sometimes be difficult to try to fit an idea for a lyrical line into the musical structure of a pre-composed song line. In some cases the integrity of clients' ideas may be better preserved if they do not have to change an idea in order to fit it into an existing rhythmic or melodic structure. Clients are then able to write lyrics freely and the music is composed to meld with the structure of the lyrics.

Musical genre and style

A good place to start the music creation is with the clients' preference for musical genre. This ground is familiar to most people, musicians and non-musicians alike. Most clients are able to state which genres of music they like or listen to, and if they cannot, the therapist can often determine this by asking the clients which artists or bands they like. It is important for the therapist to use language that the client can understand in order to promote maximum control and ownership of the music. It is often useful to ask questions about genre preference in relation to specific artists. For example, 'Do you want the music for your song to sound like a Metallica song or a Ben Harper song?'

Once the genre has been selected, different accompaniment styles can be presented for the client to choose between. These ideas may be improvised on an instrument by the therapist or trialled using an electronic keyboard accompaniment program or computer software. The choice of instrumentation is often determined by genre preference. For example, it is very difficult to create authentic sounding hip hop music using only an acoustic guitar. For this style of music, the use of computer software with recorded samples and loops and/or keyboards with pre-programmed accompaniments may be more appropriate. If a guitar is used, different accompaniment styles such as finger picking, pizzicato chords or strummed chords may be presented and different stylistic ideas, such as reggae or bossanova rhythmic patterns, power chords, standard blues riffs or use of a slide, offered as appropriate. Using a piano or electronic keyboard, arpeggios, octaves or chords may be presented as accompaniment options. The ideas for harmonic progressions are often best presented by the therapist as improvised passages, unless the client is musically proficient. Most clients will be able to identify aurally which harmonic progressions they like or think fit best with the feel of their song.

If the client is happy to sing in order to work out melody lines, then this is the next step in the song creation process. However, in our experience, even those clients who enjoy singing often don't feel confident or comfortable in using their voice to compose melody lines. The most common method we have found for melody composition is for the therapist to provide melody options for the client to listen to and choose between. The client is also encouraged to make independent decisions about the direction of the melody line; for example, 'Do you want the end of this line to go up or down?' Providing maximum opportunity for the client to contribute to the music composition ensures greater ownership of the completed song. Similar questions in terms of the range of dynamics and tempi to be used may also be asked to the client when completing the songwriting process.

Applications of the song post-recording

A recording and written transcription of the completed song is given to the client following completion of the lyrics and music. The song may be recorded on cassette tape, or for a better quality recording, a minidisc recording can be downloaded onto a computer and burnt onto a CD. These recordings serve as a record of the therapeutic songwriting process and can be used by clients to validate their emotional journey.

Some songs may be used:

- to communicate messages to loved ones
- to record positive past memories and experiences
- for self-motivation and encouragement during difficult times
- to affirm and encourage other clients who may be experiencing similar difficulties.

Sometimes when working with children, picture books including photos of children, their family, home and friends can be incorporated with the final transcription of the song so as to provide a more visual/pictorial image of the song created.

Case example: A filthy song – genre to match emotional intensity

Sam was a young man aged 19 when he was referred for music therapy. He had received a severe traumatic brain injury resulting from a train accident. He had been physically disabled as a result of his brain injury. His speech was severely dysarthric and consequently difficult to understand. His legs had both been broken in the accident, and he had high muscle tone resulting from the brain injury that caused him a lot of pain in physiotherapy. In addition to these physical issues, Sam also had severe cognitive and behavioural issues that affected his ability to participate in his rehabilitation. He was an angry young man, with limited insight into his disabilities and need for rehabilitation. Prior to music therapy intervention, his therapy sessions were often cancelled or ended prematurely due to non-compliance or aggression. As Sam's rehabilitation progress was being hindered by his negative emotional responses and behaviour, he was referred to music therapy for emotional expression and communication needs. The opportunity for self-reflection and expression through songwriting was considered appropriate as Sam needed to work through these issues in order to maintain motivation for continued therapy.

The songwriting process was not introduced into music therapy sessions until Sam had participated in music therapy for some time and rapport between the music therapist and Sam had been established. This foundation of trust and openness led Sam to feel comfortable talking with the music therapist about his feelings. This process of rapport building was further enhanced through making music with Sam which he enjoyed and to which he related. The music therapist introduced the concept of songwriting to Sam as a potential vehicle through which he could express and capture his emotional processes. Sam had a strong emotional connection to music and therefore he was responsive to this idea. He had a particular love of heavy metal music and this was the genre he wished to use for his song.

Sam's first song was written over three sessions, and evolved very much spontaneously rather than as a reorganization of the generated ideas. In brain storming, Sam's ideas were mostly phrases full of emotional intensity. Prior to his accident Sam was studying acting and his song lyrics reflected his artistic temperament. His ideas flowed quickly and remained largely unchanged in the final version of the song. As this song was very cathartic in nature, it was important to retain the ideas in their original form to capture the intensity of emotion expressed. This is the final version of his song lyrics. The name of the place where he lives has been omitted for confidentiality.

A filthy song
Chorus
I feel like shit, I hate it here
I miss my friends, wish they were near
I feel so bad, I want to walk again like I used to do
I love my dad because he believes in me.

Verse 1
I hate this place it bores me to tears
And I feel like nothing in this universe
I hate this place I'm sick of it
Want to escape from this hell hole that I'm in
I want to go home…

Chorus

Verse 2
I was riding on a train one day
Going home to X where I've lived all my life
Both my legs were broken, they were hanging out the door
Of the train, then I passed out and I don't remember anything after that
Then I woke up here in hell.

Incorporating Sam's preference for heavy metal music and the intense negative emotions that he was expressing, a range of musical ideas and options for accompaniment style were presented for him to choose between. Sam selected minor chords to start both the verse and chorus as well as major–minor chord shifts (e.g. A major to A minor) at points within the song which may have reflected the intensity of sadness and anger he was experiencing. This contrast of emotions was also represented through a change in accompaniment style. The lyrics in the chorus expressed sadness and apathy and were represented musically through a melodic and sparse vocal line. The verses expressed anger and frustration and this was represented by a change in tonality, a more driving strumming style and an increase in dynamics. The melodic range was limited to four semitones and the use of a shifting C and C# creating musical tension and lack of change. A bridge of six bars between each verse and chorus was created using previously unused chords to express a new depth of emotion. In this part, the melody line finishes on the highest pitch within the song as if to emphasize the importance of this particular line.

This first song that Sam wrote in music therapy clearly expressed many emotional responses that he was experiencing, including loss, grief, anger, boredom and love. In it he also tells his story, the story of his accident, injury and hospitalization experience. The finished song was then recorded and the final version of the song is illustrated in Figure 6.1.

Sam wrote several songs in music therapy over the course of his inpatient rehabilitation programme. His second song explored fears of life after death and past antisocial drug-taking behaviour. The next song in his process focused on the physical pain and frustration of undertaking rehabilitation, specifically physiotherapy. It also addressed issues of body image. The final song written in his process addressed issues of the future, particularly going home and the desire for a relationship. Songwriting became a medium to document his adjustment process.

Sam's songs had other therapeutic applications in that he was able to listen to the recordings of the finished songs for self-motivation and validation of his emotions.

Case example: 'Wannabe' like a Spice Girl – song parody and musical identity

Sally was almost aged 12 when she was involved as a pedestrian in a motor vehicle accident where she sustained multiple traumas, including a severe head injury. As Sally progressed through PTA she received joint music therapy and speech pathology sessions which focused on two main areas: dysarthria and language difficulties. Her cognitive difficulties included memory impairment, psychomotor slowing, impaired problem-solving, rigid thinking and impaired social judgement. Sally was unmotivated to participate in most therapy sessions. Therefore a referral for individual music therapy sessions was made to address her emotional needs including self-expression and adjustment to hospitalization.

Music therapy assessment revealed that Sally enjoyed listening to and playing music. She had previously learnt the piano and the clarinet at school and enjoyed listening to a variety of age-appropriate popular music, particularly the Spice Girls. To address the need for self-expression, songwriting was offered to Sally and she chose to use song parody as her preferred method. She was excited about the idea of producing her own songs and chose to use the music of a popular Spice Girls song 'Wannabe' to which to write her own lyrics. This demonstrates that song parody, in particular the use of current popular songs, is the preferred method of songwriting with this age group.

Sally appeared to have little difficulty in creating ideas for lyrics, but her dysarthria adversely affected her ability to articulate her intended speech

Figure 6.1 Final version of 'A filthy song'

clearly. Sally's mother was more familiar with Sally's way of speaking and was present for most sessions. Her mother was able to assist Sally if she was experiencing difficulties in articulating her thoughts and ideas. Based on clinical experience with this population, the music therapist offered Sally a selection of subject choices for her song – family, home and hospital. Sally chose to write about her experiences in hospital, which included relationships with staff members and descriptions of negative aspects of her hospital experience.

The use of song parody provided the necessary structure and predictability of melody and rhythm to aid the organization of her ideas into a lyrical format. Sally was always quick to remind the music therapist if her lyrics were not interpreted accurately. Sally was often excited during this songwriting process and often wanted to perform actions to the songs while in her wheelchair. During these times she would often need to be reminded to remain focused on the songwriting activity. These are the lyrics to the final version of Sally's song.

Naughty nurses

Verse 1
I'll tell you what I want what I really really want
I want to get out of this dumb stupid hospital
I hate medicine medicine makes me feel sicker
Except for Baclofen

Chorus
If you want to be so naughty you've got to be like Tracey
Tracey is my favourite and she is the best
Christine and Rebecca, they're my favourites too
But hospital is so smelly 'cause I think it is

Verse 2
I'll tell you what I like what I really really like
I like Sam because she does OT with me
and then there's Prue 'cause I just like her
There's no real reason I just like her

She was promised an opportunity to videotape her performance of the completed song with actions included.

Before Sally was discharged she expressed a need to write a 'going home' song. During these final therapy sessions she would often speak of being excited about leaving hospital and because this date was so close to Christmas and her twelfth birthday, there were so many family events, holiday activities and of course presents to which to look forward. This resulted in another song parody being created using yet another Spice Girls song, 'Stop Right Now'. This song spoke of her family members, events to which she was looking forward and also the desire to walk again. Both of these song parodies were video-recorded and presented in booklet form together with pictures of Sally and her family which she was then able to take home and share with family and friends.

While these were the only two songs that Sally wrote during her rehabilitation, each song parody described two significant moments in her life – her own experience of being in hospital and also preparations for the future and returning home. Song parody provided Sally with the opportunity to express a variety of emotions, thoughts and feelings which not only assisted her adjustment to hospital but also to her future as she returned home to begin a new life.

Conclusions

This chapter has primarily addressed the therapeutic rationale for introducing songwriting to TBI clients. Songwriting can facilitate the expression of emotions and assist in the movement towards emotional adjustment. Some of the more relevant and effective therapeutic techniques used when writing songs with TBI clients have been outlined. In particular, techniques such as song parody, fill-in-the-blank, song collage and the use of rhyme can be appropriate strategies to manage the cognitive deficits with which these clients may present. At the same time these techniques can be used to encourage the creation of lyrics which express feelings. In writing song lyrics, clients have opportunities to explore their own reactions to the issues addressed within the song. This is a vital step in coming to terms with the trauma that they have experienced and the short- and long-term implications of this trauma that have so adversely impacted upon their lives.

The chapter has also highlighted the importance of creating genres of music that are appealing to the client. We have particularly emphasized the attention to detail in the selection of the accompaniment style for a client's song. The use of samples, loops and pre-programmed accompaniment styles within music software programs and modern electric keyboards can be used effectively to create genre-specific effects. When using guitar accompaniment, it is important to talk with the client about how they want the song to sound and experiment with different guitar strumming and picking styles (bossanova, reggae, ballad, hard rock, etc) to create the feel and genre for the song that is desired.

Music therapists working in rehabilitation must be flexible, creative and adaptive in their therapeutic approach to songwriting. The techniques presented in this chapter need to be at the fingertips of the informed therapist in order to facilitate clear clinical decision-making about the method of presentation. This, in turn, offers maximum choice and control over the song creation process to the TBI client, in a manner appropriate to their level of need.

CHAPTER NINE

Music Therapy with the Elderly

We have seen in the previous two chapters that it is possible to build a bridge between music therapy understandings and art therapy understandings, between music therapy and medicine. Furthermore, I have proposed a way in which we can analyse how music therapists construct their clinical understandings regarding the way that patients play. This chapter continues in the same vein, addressing the pressing problem facing modern Western society of dementia in the elderly. First, the scene is set in terms of the problem of dementia within the broader health care community. Then, a music therapy understanding is presented that compares the problem as it is seen in individual therapy. We will read that music therapy has something significant to offer in terms of treatment in what is often regarded as a hopeless problem. Indeed, as we will see in the next chapter, which deals with music therapy with HIV and AIDS patients, the diagnosis of the disease and the reactions of both patients and practitioners to such a diagnosis often compound the negative aspects of the disease itself. A form of therapy that introduces some element of hope, when grounded in a reality that all parties can agree upon, is an important step forward in treating such chronic problems.

Dementia is an important source of chronic disability, leading to both spiralling health care expenditure among the elderly and a progressive disturbance of quality of life for the patient and his family. In the United States of America the cost of institutional care for patients with dementia is estimated at over $25 billion a year (Steg 1990). If 4 per cent to 5 per cent of the North American elderly population suffer from dementia, then 1.25 per cent of the total population are suffering with the problems of severe dementia. Other estimates of the same population suggest that 15 per cent of those over the age of 65 will have moderate to severe dementia with projections to 45 per cent by the age of 90 years (Odenheimer 1989). Current estimates are that over 60 per cent of those cases of dementia result from Alzheimer's disease (Kalayam and Shamoian 1990). With anticipated increases in the population of the elderly in Europe,[1] it is timely to find treatment initiatives in the Western world which will ameliorate the impact of this problem.

Dementing illnesses, or acquired cognitive disorders, have been recognized for centuries, but little progress was made in specific diagnoses until the evolution of the nosological approach to disease and early clinical descriptions of neurosyphilis and Huntington's chorea in the 1800s. Such descriptions were further supported by concurrent understandings that suggested the influence of the brain on behaviour. The first histopathological characterizations of cognitive disorders were enabled by developments in the optical microscope. Thus, Alzheimer (1907) was able to see the neuronal degeneration and senile plaques in the brain of a 55-year-old woman with progressive memory impairment, and identify the disease which today bears his name.

While cognitive impairment is evident from behaviour, and neurohistopathy can recognize neuronal degeneration, the diagnosis of Alzheimer's disease is prone to error[2] and authors differ as to the difficulty of making a precise diagnosis (Odenheimer 1989; Steg 1990). In the early stages of the disease the symptoms are difficult to distinguish from those of normal aging, a process which itself is poorly understood. To date there exist no normative established values of what cognitive impairment or memory loss are, or what neurochemical and neurophysiological changes accompany normal aging. It is therefore extremely difficult to establish criteria for determining abnormal changes from a normal population, and the researcher/clinician must in part rely upon within-the-subject designs to indicate progressive deterioration.

A second source of error in diagnosing Alzheimer's disease is that it is masked by other conditions (see Table 9.1). Principal among these conditions is that of depression, which itself can cause cognitive and behavioural disorders. It is estimated that 20 per cent to 30 per cent of patients with Alzheimer's disease will have an accompanying depression (Kalayam and Shamoian 1990), thereby compounding diagnostic problems further.

---

1   Between 23 per cent and 25 per cent of the national populations aged over 65 by the year 2040 (Aldridge 1990).
2   Estimated in a range from 10 per cent to 30 per cent error in the general medical population (Steg 1990).

Table 9.1 Differential diagnosis of Alzheimer's disease

Differential diagnosis of Alzheimer's disease

Multi infarct dementia and other forms of cerebrovascular disease
Parkinson's disease
Progressive supranuclear palsy
Huntington's disease
Central nervous system infection
Sudbural haematoma
Normal pressure hydrocephalus
Multiple sclerosis
Seizure disorder
Brain tumour
Cerebral trauma
Metabolic disturbance
Nutritional deficiency
Psychiatric disorder
Substance abuse or overmedication

Clinical Descriptions of Dementia

The clinical syndrome of dementia is characterized by an acquired decline of cognitive function which is represented by memory and language impairment. While the term dementia itself is used widely to describe cognitive impairment it is specifically applied in medical literature to two conditions: dementia of the Alzheimer's type (DAT) and multi-infarct dementia.

The course of Alzheimer's disease is one of progressive deterioration associated with degenerative changes in the brain. Such deterioration is presented in a clinical picture of episodic changes and a pattern of particular cognitive failings which are variable (Drachman *et al.* 1990). Mental status testing is one of the primary forms of assessing these cognitive failings, which include short- and long-term memory changes, impairment of abstract thinking and judgement; disorders of language (aphasia), and difficulty in finding the names of words (anomia); the loss of ability to interpret what is heard, said and felt (agnosia); and an inability to carry out motor activities, such as manipulating a pen or toothbrush, despite intact motor function (apraxia). When such clinical findings are present then a probable diagnosis can be made; a more definite diagnosis depends upon tissue diagnosis (see Table 9.2).

Table 9.2 Diagnostic evaluation of dementia

Diagnostic categories

Complete medical history
Mental status examination
Complete physical and neurological investigation (including investigation for infection of central nervous system if suspected)
Complete blood count and blood chemistry tests (including vitamin B12 levels)
Thyroid function tests
Serology for syphilis
Computerized tomography (CT) or magnetic resonance imaging (MRI), electroencephalography (EEG), or positive emission tomography (PET) scanning

While dementia of the Alzheimer's type begins after the age of 40, and is considered to be a disease of the elderly, the influence of age on prognosis is not as significant as the initial degree of severity of the problem when recognized (Drachman *et al.* 1990). Disease severity, as assessed by intellectual function, appears to be the most consistent predictor of the subsequent course of the disease, particularly when accompanied by a combination of wandering and falling, and behavioural problems (Walsh, Welch and Larson 1990). However, the rates of decline between sub groups of patients are variable and a patient's rate of progression in one year may bear little relationship to the future rate of decline (Salmon *et al.* 1990). Some authors (Cooper, Mungas and Weiler 1990) suggest that an as yet unproven factor, other than declining cognitive ability, may also play a part in the associated abnormal behaviours of anger, agitation, personality change, wandering, insomnia and depression which occur in later stages of the disease.

Clearly Alzheimer's disease causes distress for the patient. The loss of memory and the accompanying loss of language, before the onset of motor impairment, mean that the daily lives of patients are disturbed. Communication, the fabric of social contact, is interrupted and disordered. The threat of progressive deterioration and behavioural disturbance has ramifications not only for the patient themselves, but also their families, who must take some of the social responsibility for care of the patient, and the emotional burden of seeing a loved one becoming confused and isolated.

Assessment of dementia

A brief cognitive test, the Mini-Mental State Examination (Folstein, Folstein and McHugh 1975), has been developed to screen and monitor the progression of Alzheimer's disease. The test itself is intended for the clinician to assess functions of different areas of the brain, and is based upon questions and activities (see Table 9.3). As a clinical instrument it is widely used and well validated in practice (Babikian *et al.* 1990; Beatty and Goodkin 1990; Eustache *et al.* 1990; Faustman, Moses and Csernansky 1990; Gagnon *et al.* 1990; Jairath and Campbell 1990; Summers *et al.* 1990; Zillmer *et al.* 1990). As a bedside test the MMSE is widely used for testing cognition and is useful as a predictive tool for cognitive impairment and semantic memory (Eustache *et al.* 1990) without being contaminated by motor and sensory deficits (Beatty and Goodkin 1990; Jairath and Campbell 1990).

Table 9.3 Mini-Mental State Examination

Item	Component	Score
Orientation for time	year, season, month, date and day	5
Orientation for place	state, county, city, building and floor	5
Registration	Subject repeats 'rose', 'ball' and 'key'	3
Attention for calculation	Serial subtraction of 7 from 100 or spell 'world' backward	5
Recall	'Rose', 'ball' and 'key'	3
Naming	Pencil and watch	2
Repetition	No ifs, ands, or buts	1
Three-stage verbal command	Take a piece of paper in your right hand, fold it in half, and put in on the floor	3
Written command	Close your eyes	1
Writing	A spontaneous sentence	1
Construction	Two interlocking pentagons	1
Total		30

Source: after Folstein, Folstein and McHugh (1975).

Elderly patients scoring below 24 points out of a possible total score of 30 are considered as demented. However, this scoring has been questioned on the grounds of its cut-off point of 24 as the lower limit, particularly for early dementia (Galasko *et al.* 1990); and that it is influenced by education (Gagnon *et al.* 1990). Poorly educated subjects with less than eight years of education may score below 24 without being demented.

Further criticisms of the Mini-Mental State Examination (MMSE) have been that it is not sensitive enough to mild deficits, but it could be augmented by the addition of a word fluency task and an improvement in the attention–concentration item (Galasko *et al.* 1990). In addition, the MMSE seriously underestimates cognitive impairment in psychiatric patients (Faustman, Moses and Csernansky 1990). An important feature neglected by the MMSE is that of 'intention' or executive control (Odenheimer 1989), which refers to the ability of the patient to persevere with a set task, to reach a set goal or to change tasks.

The items which the MMSE fails to discriminate (minor language deficits), or neglects to assess (fluency and intentionality), however, may be elicited in the playing of improvised music. A dynamic musical assessment of patient behaviour, linked with the motor co-ordination and intent required for the playing of musical instruments used in music therapy, and the necessary element of interpersonal communication, may provide a sensitive complementary tool for assessment.

We see in Table 9.4 how medical elements of assessment can find their correlates in musical parameters. As we have seen in the chapter related to bowel disease, both languages share similar terms, and it is possible to build a conceptual bridge between two forms of practice. By doing this neither practice is reduced to the other, but we do have a valuable conceptual tool for proposing commonalities in the practice. We can take all the elements demanded in a medical assessment and translate those into terms that are applicable for music therapy. Conversely, we can translate what happens in music therapy and demonstrate its applicability to medical practice. What is important to take from the music therapy assessment is that the idea of intention (which is behavioural as much as it is cognitive) can be demonstrated in an activity that offers a situation in which intentionality can be achieved. The problem with assessment tests is that they are so often reduced that they become unnatural and totally divorced from the context of the person's life. How we behave in laboratories and consulting rooms is often somewhat different from our behaviour in our own homes and with friends. Music therapy, too, is no natural context; but embedding an assessment of intentionality within the context of musical playing is less clumsy, and more flexible, than a specific test.

understand speech he was no longer capable of the co-ordination required to lead a major orchestra. While his mind, he reports, was full of musical ideas, he could not set them down (Dalessio 1984). Eventually his intellectual functions and speech deteriorated until he could no longer recognize his own music.

However, the responsiveness of patients with Alzheimer's disease to music is a remarkable phenomenon (Swartz et al. 1989). While language deterioration is a feature of cognitive deficit, musical abilities appear to be preserved. This may be because the fundamentals of language, as we have seen in previous chapters, are musical, and prior to semantic and lexical functions in language development.

Although language processing may be dominant in one hemisphere of the brain, music production involves an understanding of the interaction of both cerebral hemispheres. (Altenmüller 1986; Brust 1980; Gates and Bradshaw 1977). In attempting to understand the perception of music there have been a number of investigations into the hemispheric strategies involved. Much of the literature considering musical perception concentrates on the significance of hemispheric dominance. Gates and Bradshaw (1977) conclude that cerebral hemispheres are concerned with music perception and that no laterality differences are apparent. Other authors (Wagner and Hannon 1981) suggest that two processing functions develop with training where left and right hemispheres are simultaneously involved, and that musical stimuli are capable of eliciting both right and left ear superiority (Kellar and Bever 1980). Similarly, when people listen to and perform music they utilize differing hemispheric processing strategies.

Evidence of the global strategy of music processing in the brain is found in the clinical literature. Morgan and Tilluckdharry (1982) tell how singing was considered to be a welcome release from the helplessness of being a patient, allowing thoughts to be communicated externally. Although the 'newer aspect' of speech was lost, the older function of music was retained. Berman (1981) suggests that recovery from aphasia is not a matter of new learning by the non-dominant hemisphere but a taking over of responsibility for language by that hemisphere. The non-dominant hemisphere may be a reserve of functions in case of regional failure. If singing gives a glimpse of this brain plasticity, then music therapy has a potential for working with dementia patients.

Little is known about the loss of musical and language abilities in cases of global cortical damage, although the quality of response to music in the final stages of dementia is worth noting (Norberg, Melin and Asplund 1986). Any discussion is necessarily limited to hypothesizing, as there are no

Table 9.4 Features of medical and musical assessment

Medical elements of assessment	Musical elements of assessment
continuing observation of mental and functional status	continuing observation of mental and functional status
testing of verbal skills, including element of speech fluency	testing of musical skills; rhythm, melody, harmony, dynamic, phrasing, articulation
cortical disorder testing; visuo-spatial skills and ability to perform complex motor tasks (including grip and right–left co-ordination)	cortical disorder testing; visuo spatial skills and ability to perform complex motor tasks (including grip and right–left co-ordination)
testing for progressive memory disintegration	testing for progressive memory disintegration
motivation to complete tests, to achieve set goals and persevere in set tasks	motivation to sustain playing improvised music, to achieve musical goals and persevere in maintaining form
'intention' difficult to assess; but considered important	'intention' a feature of improvised musical playing
concentration and attention span	concentration on the improvised playing and attention to the instruments
flexibility in task switching	flexibility in musical (including instrumental) changes
mini-mental state score influenced by educational status	ability to play improvised music influenced by previous musical training
insensitive to small changes	sensitive to small changes
ability to interpret surroundings	ability to interpret musical context and assessment of communication in the therapeutic relationship

Music and Dementia

Late in adult life, at the age of 56, and after completing two major concertos for the piano, Maurice Ravel, the composer, began to complain of increased fatigue and lassitude. Following a traffic accident his condition deteriorated progressively (Henson 1988). He lost the ability to remember names, to speak spontaneously and to write (Dalessio 1984). Although he could established baselines for musical performance in the adult population (Swartz et al. 1989). Aphasia, which is a feature of cognitive deterioration, is a complicated phenomenon. While syntactical functions may remain longer, it is the lexical and semantic functions of naming and reference which begin to fail in the early stages. Phrasing and grammatical structures remain, giving an impression of normal speech, yet content becomes increasingly incoherent. These progressive failings appear to be located within the context of semantic and episodic memory loss illustrated by the inability to remember a simple story when tested (Bayles et al. 1989).

Musicality and singing

Musicality and singing are rarely tested as features of cognitive deterioration, yet preservation of these abilities in aphasics has been linked to eventual recovery (Jacome 1984; Morgan and Tilluckdharry 1982), and could be significant indicators of hierarchical changes in cognitive functioning. Jacome (1984) found that a musically naive patient with transcortical mixed aphasia exhibited repetitive, spontaneous whistling, and whistling in response to questions. The patient often spontaneously sang without error in pitch, melody, rhythm and lyrics, and spent long periods of time listening to music. Beatty (Beatty et al. 1988) describes a woman who had severe impairments in terms of aphasia, memory dysfunction and apraxia, yet was able to sight read an unfamiliar song and perform on the xylophone which to her was an unconventional instrument. Like Ravel (Dalessio 1984), and an elderly musician who could play from memory (Crystal, Grober and Masur 1989) but no longer recalled the name of the composer, she no longer recalled the name of the music she was playing.

Swartz and his colleagues (Swartz et al. 1989, p.154) propose a series of perceptual levels at which musical disorders take place:

(1) the acoustico-psychological level, which includes changes in intensity, pitch and timbre

(2) the discriminatory level, which includes the discrimination of intervals and chords

(3) the categorical level which includes the categorical identification of rhythmic patterns and intervals

(4) the configural level, which includes melody perception, the recognition of motifs and themes, tonal changes, identification of instruments, and rhythmic discrimination; and

(5) the level where musical form is recognized, including complex perceptual and executive functions of harmonic, melodic and rhythmical transformations.

In Alzheimer's patients it would be expected that while levels (1), (2) and (3) remain unaffected, the complexities of levels (4) and (5), when requiring no naming, may be preserved but are susceptible to deterioration.

It is perhaps important to point out that these disorders are not themselves musical; they are disorders of audition. Only when disorders of musical production take place can we begin to suggest that a musical disorder is present. Improvised musical playing is in a unique position to demonstrate this hypothetical link between perception and production.

Rhythm is the key to the integrative process, underlying both musical perception and physiological coherence. Barfield's (1978) approach suggests that when musical form as tonal shape meets the rhythm of breathing there is the musical experience. External auditory activity is mediated by internal perceptual shaping in the context of a personal rhythm. When considering communication, rhythm is also fundamental to the organization and co-ordination of internal processes, and externally between persons (Aldridge 1989, 1989a, 1991d).

Music Therapy and the Elderly

Much of the published work concerning music therapy with the elderly is concerned with group activity (Bryant 1991; Christie 1992; Olderog Millard and Smith 1989) and is generally used to expand socialization and communication skills, with the intention of reducing problems of social isolation and withdrawal, to encourage participants to interact purposefully with others, assist in expressing and communicating feelings and ideas, and to stimulate cognitive processes, thereby sharpening problem-solving skills. Additional goals also focus on sensory and muscular stimulation and gross and fine motor skill preservation (Segal 1990).

Clair (1990, 1991; Clair and Bernstein 1990, 1990a) has worked extensively with the elderly and found music therapy a valuable tool for working in groups to promote communicating, watching others, singing, interacting with an instrument, and sitting. Her main conclusions are that although the group members deteriorated markedly in cognitive, physical, and social capacities over an observation period of 15 months, they continued to participate in music activities. During the 30-minute sessions the group members consistently sat in chairs without physical restraints for the duration of each session, and interacted with others regardless of their deterioration. This was the only time in the week when they interacted with others (Clair and Bernstein 1990a). Indeed, for one 66-year-old man was it the sensory stimulation of music therapy that brought him out of his isolation such that he could participate with others, even if for a short while (Clair 1991).

Wandering, confusion and agitation are linked problems common to elderly patients living in hostels or special accommodation for Alzheimer's patients. A music therapist (Fitzgerald Cloutier 1993) has tested singing with an 81-year-old woman to see if it helped her to remain seated. After 20 singing sessions, the therapist read to the woman to compare the degree of attentiveness. While music therapy and reading sessions redirected the woman from wandering, the total time she sat for the music therapy sessions was double that of the reading sessions (214.3 mins vs 99.1 mins), and the time spent seated in the music therapy was more consistent than the episodes when she was being read to. When agitation occurs in such elderly women, then individualized music therapy appears to have a significantly calming effect (Gerdner and Swanson 1993). In terms of reducing repetitive behaviour, musical activity also reduces disruptive vocalizations (Casby and Holm 1994).

The above conclusions are supported by Groene (1993). Thirty residents (aged 60–91 years) of a special Alzheimer's unit, who exhibited wandering behaviour, were randomly assigned either to mostly music attention or to mostly reading attention groups, where they received one-to-one attention. Those receiving music therapy remained seated longer than those in the reading sessions.

One of the central problems of the elderly is the loss of independence and self-esteem, and Palmer (1983, 1989) describes a programme of music therapy at a geriatric home designed to rebuild self-concept. For the 380 residents, ranging from those who were totally functional to those who needed total care, a programme was adapted to the capacities and needs of individual patients. Marching and dancing increased the ability of some patients to walk well; and for the non-ambulatory, kicking and stamping to music improved circulation and increased tolerance and strength. Sing-along sessions were used to encourage memory recall and promoted social interaction and appropriate social behaviour. (Palmer 1983, 1989). It was such social behaviour that Pollack (Pollack and Namazi 1992) reports as being accessible to improvement through group music therapy activities. It is the participative element that appears to be valuable for communication, and the intention to participate that is at the core of the music therapy activity, which we will see in the following section.

Music therapy has also been used to focus on memory recall for songs and the spoken word (Prickett and Moore 1991). In ten elderly patients, whose diagnosis was probably Alzheimer's disease, words to songs were recalled dramatically better than spoken words or spoken information. Although long-familiar songs were recalled with greater accuracy than a newly presented song, most patients attempted to sing, hum, or keep time while the therapist sang. However, Smith (1991) suggests that it is factors such as tempo, length of seconds per word, and total number of words that might be more closely associated with lyric recall than the relative familiarity of the song selection.

In a further study of the effects of three treatment approaches, musically cued reminiscence, verbally cued reminiscence, and music alone, on the cognitive functioning of 12 female nursing home residents with Alzheimer's disease, changes in cognitive functioning were assessed by the differences between pre- and post-session treatment scores on the Mini-Mental State Examination. Comparisons were made for total scores and sub-scores for orientation, attention, and language. Musically cued and verbally cued reminiscence significantly increased language sub-section scores and musical activity alone significantly increased total scores (Smith 1986).

Music therapy with an Alzheimer's patient

Nordoff–Robbins music therapy is based upon the improvisation of music between therapist and patient (Nordoff and Robbins 1977). The music therapist plays the piano, improvising with the patient, who uses a range of instruments. This work often begins with an exploratory session using rhythmic instruments, in particular the drum and cymbal; progressing to the use of rhythmic/melodic instruments such as the chime bars, glockenspiel or xylophone; developing into work with melodic instruments (including the piano); and the voice. In this way of working the emphasis is on a series of musical improvisations during each session, and music is the vehicle for the therapy. Each session is audiotape-recorded, with the consent of the patient, and later analyzed and indexed as to musical content.

In the case example below music therapy is used as one modality of a comprehensive treatment package. The patient is seen on an outpatient basis for ten weekly sessions. Each session lasts for 40 minutes. She is unable to find her way on public transport and is brought to the hospital by her son.

Edith was a 55-year-old woman who came to the hospital for treatment. Her sister had died of Alzheimer's disease, and the family were concerned that she too was repeating her sister's demise as her memory became increasingly disturbed. She began playing the piano for family, friends and acquaintances at the age of 40, although without any formal studies, and, given this interest, music therapy appeared to have potential as an intervention adjuvant to medical treatment.

The patient was referred initially to the hospital when she, and her son, became aware of her own deteriorating condition, although the disease was in its early stages. At home she was experiencing difficulties in finding clothing and other things necessary for everday life. She could not cook for herself anymore and was unable to write her own signature. While wanting to speak, she experienced difficulty in finding words. It may be assumed that given the family background and her own understanding of her failings, the cognitive problem was exacerbated by depression.

Characteristics of the musical playing
RHYTHMIC PLAYING

In all ten sessions Edith demonstrated her ability to play, without the influence of her music therapist, a singular ordered rhythmical pattern in 4/4 time using two sticks on a single drum. This rhythmical pattern appeared in various forms and can be portrayed as seen in Figure 9.1. Example 1.

A feature of her rhythmical playing was that in nearly all the sessions, during the progress of an improvisation, the patient would let control of the rhythmic pattern slip such that it became progressively imprecise, losing both its form and liveliness. The initial impulse of her rhythmical playing, which was clear and precise, gradually deteriorated as she lost concentration and ability to persevere with the task in hand. However, when the therapist offered an overall musical structure during the course of the improvisation, the patient could regain her precision of rhythm. It could well be that to sustain perception an overall rhythmical structure is necessary, and it is this musical gestalt, that is, the possibility of providing an overall organizing structure of time, which fails in Alzheimer's disease.

The patient reacted quickly to changes in time and different rhythmic forms, and incorporated these within her playing. Significantly, she reacted fluently in her playing to changes from 4/4 time to 3/4 time, often remarking, 'Now it's a waltz...'. With typical well-known rhythmical forms, e.g. the Habaner rhythm, in combination with characteristic melodic phrases, she laughed, breathed deeply and played with stronger, more thoughtful intent.

These rhythmical improvisations, using different drums and cymbals, were played in later sessions on two instruments together. The patient had no difficulty in controlling and maintaining her grip of the beaters. Similarly she showed no difficulty in co-ordinating parallel or alternate handed

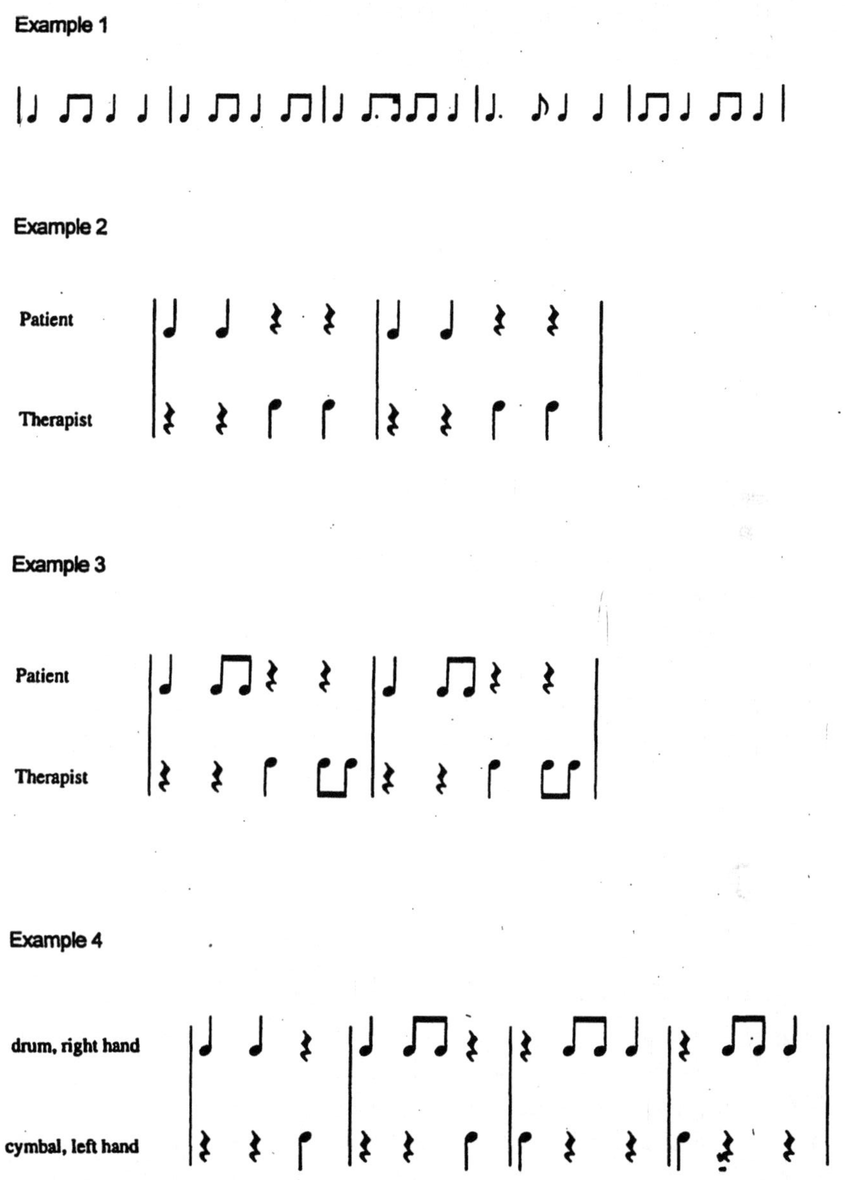

Example 1

Example 2
Patient
Therapist

Example 3
Patient
Therapist

Example 4
drum, right hand
cymbal, left hand

Figure 9.1 Examples of musical playing

playing on a single instrument, although she played mostly with a quick tempo (120 beats per minute). However, the introduction of two instruments brought a major difficulty for the patient, who stood disorientated before the instruments, unable to integrate them both in the playing. It was only with instructions and direction from the therapist that the patient was able to co-ordinate right–left playing on two instruments; changes in the pattern of the playing were also difficult to realize (see Figure 9.1, Examples 2 and 3).

Throughout the improvisations the inherent musical ability of the patient, in terms of tempo (ritardando, accelerando, rubato) and dynamic (loud and soft), was expressed whenever she had the opportunity to.

MELODIC PLAYING

Melody is an expression of motion which arises and decays from moment to moment. In this motion the size of the intervals provides a melodic tension which itself has a dynamic power. The experience of melody is itself an experience of form. As a melody begins there is the possibility of grasping a sense of the immediacy of the whole form, and preparing for the aesthetic pleasure of deviations from what is expected. This element of tension between the expected and the unpredictable has been at the heart of musical composition for the last two hundred years. In addition, it is melody that leads the music from the rhythmical world of feeling into the cognitive world of imagination.

When Edith played her melodies were always lively. She knew many folk songs from earlier times and was able to sing them alone. After only a few notes played by the therapist on the piano she could associate those notes with a well known tune. However, when she tried to play a complete melody on the piano or other melody instrument, it proved impossible. Although beginning spontaneously and fluently she had difficulty in completing a known melody.

Melody instruments like the metallophone and the xylophone, which were previously unknown to the patient, remained forever strange. At the introduction of a new melody she would often seek a previously known melody rather than face the insecurity of an unknown improvised melody. When the therapist sat opposite her and showed her which notes to play she was able to follow the therapist's finger movements. When presented with a limited range of tones she also had difficulty in playing them, which may have been compounded by visuo-spatial difficulties, in that it is easier to strike the surface of a drum than the limited precise surfaces of adjacent chime bars.

At the beginning of the very first session after entering the therapy room, Edith set her eyes on the piano and began spontaneously to play 'Happy is the Gypsy Life'. She easily accompanied this song harmonically with triads and thirds. The second song which she attempted to play proved more difficult as she failed to find the subdominant, whereupon she broke off from the playing and remarked '…that always catches me out'. This pattern of spontaneously striking up a melody and then breaking off when the harmony failed was to be repeated whenever she tried other songs like 'Happy Birthday' and 'Horch was kommt von draußen rein'. She showed a fine musical sensitivity for the appropriate harmony, which was not always at her disposal to be played. In the playing of the drum her musical sensitivity in her reactions to the contrasting sound qualities of major and minor was reduced, but overall she had a pronounced perception of this harmonic realm of music. We see here, as in tests of language functioning, that the production, in this case of music, is impaired while perception remains.

Changes in the musical playing of the patient

In the rhythmical playing on drum and cymbal the therapist attempted to develop the patient's attention span through the use of short repeated musical patterns and changes in key, volume and tempo. She hoped to through changes in the sound to steer the patient to maintaining a stable musical form. This technique helped the patient to maintain a rhythmical pattern and brought her to the stage in which she could express her self, more strongly, musically. Beyond the emphasis on the basic beat in the music, the therapist searched for other ways to respond to, and develop a variety in, rhythmical pattern by moving away from the repetitive pattern played by the patient. In a quick tempo the patient was able to maintain a basic beat for a certain time. As soon as the tempo changed and became slower, or the music varied with the introduction of a semiquaver, the stable element of the music was disturbed and took on a superficial character.

A further change in the improvising was shown when the patient recognized, and could repeat rhythmical patterns which were frequently realized as a musical dialogue and brought into a musical context. In the last session of therapy the patient was able to change her playing in this way such that she could express herself more strongly by bringing her thoughtful and expressive playing into line (see Figure 9.1, Example 4).

A crucial point in the music came when she chose to play for a bar on the cymbal. Although after a while she trusted herself to play without help on two instruments, she could not get to grips with a new personal initiative on these instruments. This was also reflected in her continuing difficulty with what were initially strange instruments, like the temple blocks. She also expressed her insecurity as to how to proceed, and needed instructions about what to do next.

She displayed few changes in her dynamic playing. She reacted to dynamic contrasts and transitions, but powerful *forte* playing was only achieved in the last session. At times her playing had a uniform quality of attack which gave it a mechanistic and immovable character. For this patient it was not possible to build a freely improvised melody from a selection of tones. It was as if she was a prisoner of the search for melodies of known fixed songs; therefore she chose the free form of improvising on rhythm instruments.

INTENTIONAL PLAYING

From the first session of therapy the patient made quite clear her intent to sit at the piano, and play whatever melodies she chose and find the appropriate accompaniments. This wish and the corresponding will-power to achieve this end were shown in all the sessions. It was possible to use this impetus to play as a source for improvisation. In the sixth session Edith improvised a rhythmical piece in 4/4 time, which the therapist then transformed with a melodic phrase. At the end of the phrase the patient laughed with joy at the success of her playing and asked to play it again. The original lapses and slips in the form of the rhythmical playing could be carried by the intent and expression with which she played. While her overall intention to play was preserved, her attention to that playing, the concentration necessary for musical production and the perseverance required for completing a sequence of phrases progressively failed, and were dependent on the overall musical structure offered by the therapist.

Clinical changes

At the end of the treatment period, which also used homeopathic medicine, she was able to cook for herself and could find her own things about the house. The psychiatrist responsible for her therapeutic management reported an overall improvement in her interest in what was going on around her, and in particular that she maintained attention to visitors and conversations. The patient regained the ability to write her signature, although she could only write slowly. While wanting to speak, she still experienced difficulty in finding words.

It appears that music therapy has a beneficial effect on the quality of life of this patient, and that some of the therapeutic effect may have been brought about by treating the depression. While the patient came to the sessions with the intention of playing her ability to take initiatives was impaired, mirroring the state of her home life, where she wanted to look after herself yet was unable to take initiatives. The stimulus to take initiatives was seen as an important feature of the music therapy by the therapist, and appears to have a correlate in the way in which the patient began to take initiatives in her daily life. Active music making also promotes interaction between the persons involved, thereby promoting initiatives in communication which the patient also enjoyed, particularly when she accomplished a complete improvisation. Furthermore, the implications for the maintenance of memory by actively making music is significant. As Crystal *et al.* (1989) found in an 82-year-old musician with Alzheimer's disease, the ability to play previously learned piano compositions from memory was preserved, although the man was unable to identify the composer or titles of each work; and the ability to learn the new skill of mirror reading was also preserved while the man was unable to recall or recognize new information.

A contra-indication for music therapy with such patients, who are aware of their problems, is that the awareness of further deterioration in cognitive abilities (as this patient experienced in her piano playing) may exacerbate any underlying depression and demotivate the patient to continue. For Edith, not being able to find the appropriate harmonies to well-known tunes, that she could play when she was younger was yet another sign of her deteriorating status.

Music Therapy as a Sensitive Tool for Assessment

If we are unsure as to the normal process of cognitive loss in aging, we are even more in the dark as to the normal musical playing abilities of adults. The literature suggests that musical activities are preserved while other cognitive functions fail. Alzheimers patients, despite aphasia and memory loss, continue to sing old songs and to dance to past tunes when given the chance. Indeed, fun and entertainment are all part and parcel of daily living for the elderly living in special accommodation (Glassman 1983; Jonas 1991; Kartman 1990; Smith 1992). Quality of life expectations become paramount in any management strategy, and music therapy appears to play an important role in enhancing the ability to take part actively in daily life (Lipe 1991; Rosling and Kitchen 1992). However, the production and improvisation of music appear to fail in the same way in which language fails. Unfortunately no established guidelines as to the normal range of improvised music playing of adults is available.

Improvised music therapy appears to offer the opportunity to supplement mental state examinations in areas where those examinations are lacking (see

Tables 9.4 and 9.5). First, it is possible to ascertain the fluency of musical production. Second, intentionality, attention to, concentration on and perseverance with the task in hand are important features of producing musical improvisations and susceptible to being heard in the musical playing. Third, episodic memory can be tested in the ability to repeat short rhythmic and melodic phrases. The inability to build such phrases may be attributed to problems with memory or to an as yet unknown factor. This unknown factor is possibly involved with the organization of time structures. If rhythmic structure is an overall context for musical production, and the ground structure for perception, it can be hypothesized that it is this overarching structure which begins to fail in Alzheimer's patients. A loss of rhythmical context would explain why patients are able to produce and persevere with rhythmic and melodic playing when offered an overall structure by the therapist. Such a hypothesis would tie in with the musical hierarchy proposed by Swartz (Swartz et al. 1989, p.154), and would suggest a global failing in cognition while localized lower abilities are retained. However, the hierarchy of musical perceptual levels proposed by Swartz may need to be further sub-divided into classifications of music reception and music production.

Table 9.5 Musical elements of assessment and examples of improvised playing

Musical elements of assessment	Examples of improvised playing
testing of musical skills; rhythm, melody, harmony, dynamic, phrasing, articulation	improvisation using rhythmic instruments (drum and cymbal) singly or in combination
	improvisation using melodic instruments
	singing and playing folk songs with harmonic accompaniment
cortical disorder testing: visuo-spatial skills	playing tuned percussion (metallophone, xylophone, chime bars) demanding precise movements
cortical disorder testing: ability to perform complex motor tasks (including grip and right–left co-ordination	alternate playing of cymbal and drum using a beater in each hand
	co-ordinated playing of cymbal and drum using a beater in each hand
	co-ordinated playing of tuned percussion

Table 9.5 Musical elements of assessment and examples of improvised playing (continued)

Musical elements of assessment	Examples of improvised playing
testing for progressive memory disintegration	the playing of short rhythmic and melodic phrases within the session, and in successive sessions
motivation to sustain playing improvised music, to achieve musical goals and preserve in maintaining musical form	the playing of a rhythmic pattern deteriorates when unaccompanied by the therapist as does the ability to complete a known melody, although tempo remains
'intention' a feature of improvised musical playing	the patient exhibits the intention to play the piano from the onset of therapy and maintains this intent throughout the course of treatment
concentration on the improvised playing and attention to the instruments	the patient loses concentration when playing, with qualitative loss in the musical playing and lack of precision in the beating of rhythmical instruments
flexibility in musical (including instrumental) changes	initially the musical playing is limited to a tempo of 120 Bp and a characteristic pattern but this is responsive to change
ability to play improvised music influenced by previous musical training	although the patient has a musical background this is only of help when she perceives the musical playing; it is little influence in the improvised playing
sensitive to small changes	musical changes in tempo, dynamic, timbre and articulation which at first are missing are gradually developed
ability to interpret musical context and relationship	the patient develops the ability to play in a musical dialogue with the therapist demanding both a refined musical perception and the ability of musical production

Music therapy appears to offer a sensitive assessment tool (see Table 9.5). It tests those prosodic elements of speech production which are not lexically dependent. Furthermore, it can be used to assess those areas of functioning, both receptive and productive, not covered adequately by other test instruments; i.e. fluency, perseverance in context, attention, concentration and intentionality. In addition it provides a form of therapy which may stimulate cognitive activities such that areas subject to progressive failure are maintained. Certainly the anecdotal evidence suggests that quality of life of Alzheimer's patients is significantly improved with music therapy (McCloskey 1985, 1990; Tyson 1989) accompanied by the overall social benefits of acceptance and sense of belonging gained by communicating with others (Morris 1986; Segal 1990).

Prinsley recommends music therapy for geriatric care as it reduces the individual prescription of tranquilizing medication, reduces the use of hypnotics on the hospital ward, and helps overall rehabilitation. He recommends that music therapy be based on treatment objectives: the social goals of interaction co-operation; psychological goals of mood improvement and self-expression; intellectual goals of the stimulation of speech and organization of mental processes; and the physical goals of sensory stimulation and motor integration (Prinsley 1986). Such goals as stimulation of the individual, promoting involvement in social activity, identifying specific individualized behavioural targets, and emphasizing the maintenance of specific memory functions are repeated throughout the music therapy literature (Prange 1990; Smith 1990; Summer 1981). Similarly, Smith (1990a) recommends behavioural interventions targeted at the more common behavioural problems (e.g., disorientation, age-related changes in social activity, sleep disturbances) of institutionalized elderly persons. In a matched control study of music therapy, or no music therapy, in two nursing homes, life satisfaction and self-esteem were significantly improved in the home where the residents participated in the musical activities (VanderArk, Newman and Bell 1983).

In terms of research, single-case within-subject designs with Alzheimer's appear to be a feasible way forward to assess individual responses to musical interventions in the clinical realm. Such studies would depend upon careful clinical examinations, mental state examinations and musical assessments. Unfortunately most of the literature concerning cognition and musical perception is based upon audition and not musical production. Like other authors we suggest that the production of music, as is the production of language, is a complex global phenomenon as yet poorly understood. The understanding of musical production may well offer a clue to the ground structure of language and communication in general. It is research in this realm of perception which is urgent not only for the understanding of Alzheimer's patients but in the general context of cognitive deficit and brain behaviour. It may be, as Berman (1981) suggests, that the non-dominant hemisphere is a reserve of functions in case of regional failure and this functionality can be stimulated to delay the progression of degenerative disease. Furthermore, it is important to point out that when the overall rhythmic pattern failed, the patient was able to maintain her beating in tempo. A similar situation applies in coma patients who cannot co-ordinate basic life pulses within a rhythmic context and thereby regain consciousness (Aldridge 1991b; Aldridge, Gustorff and Hannich 1990). We may need to address in future research the co-ordinating role of rhythm in human cognition and consciousness, whether it be in persons who are losing cognitive abilities, or in persons who are attempting to gain cognitive abilities.

ALZHEIMER'S DISEASE

rhythmic playing

loss of precision	⊃ loss of concentration and intent	→ sign of progressive deterioration
loss of precision	⊃ loss of musical structure	→ perception unsustained
patient plays melody rhythm	⊃ regains form	→ plays with intent

ALZHEIMER'S DISEASE

melodic playing

melodic tension	⊃ dynamic power	→ experience of form
patients plays melody	⊃ moves from world of feeling	→ enters world of cognition

Figure 9.2 Constitutive rules relating to the rhythmic and melodic improvised playing of

As a research point we can see how the use of texts relating to clinical practice can be used to build up a picture of clinical reasoning, as we also saw at the end of the last chapter. In this case we have constitutive rules (see Figure 9.2) relating to the rhythmical playing, where rhythmical precision is seen as a sign of progressive deterioration and evidence of a loss of concentration or loss of intent. Conversely, the maintenance of rhythm is construed as regaining form and evidence of playing with intent.

It is not only texts themselves that can be used to elicit meanings. We can see graphically in Figure 9.3 how a 'musical text' can be used as the basis for therapeutic assessment using a constitutive rules perspective. This assessment is based upon the same musical material as seen in Figure 9.1, Example 1 and the constitutive rule seen in Figure 9.2. By presenting material in this way we have a flexible means by which musical understandings can be constructed and demonstrated. That musical texts can be used is an important bridge between the experiences of the musician as therapist and the musician as researcher. This is a vertical understanding where level 1 is the musical experience itself, that is then 'neutrally' described at level 2, and interpreted at level 3.

Using these elements of rhythmical playing we can construct regulative, horizontal rules as they refer to the therapy process in time. In Figure 9.4 we see how when a repetitive pattern, which is interpreted as an inability on the part of the patient to express herself, is played, the therapist varies her own rhythmical pattern. If the patient recognizes such a change in rhythmical pattern, she is able to express herself communicatively and maintain the therapeutic dialogue. Having both a vertical form of analysis and a horizontal form of analysis allows us in some ways to display the element of performance that we have in a musical score. Vertically we have harmony and horizontally we have a substrate of time. In my method we have a vertical complexity of elicited understandings that are constructed horizontally as descriptions of therapeutic development. The benefit of this approach is that we can be clear about the material we are using as a basis for our description, and elucidate the stages of interpretation as they move away from the experience itself.

RHYTHMIC PLAYING OF A PATIENT WITH ALZHEIMER'S DISEASE

therapeutic assessment

♩ ♫ ♫ | ♩ ♬ ♬ | ♩. ♪ ♩ ♩ | ⊃	loss of precision	→	sign of progressive deterioration
Level 1 musical event	Level 2 description		Level 3 interpretation

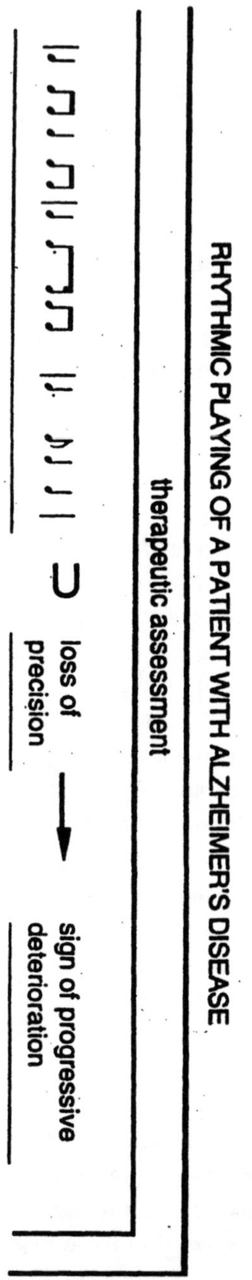

Figure 9.3 Textual understandings of a musical phenomenon in therapeutic assessment

context of close friendship

[patient plays a repetitive pattern] ⊃ { invalid: loses musical contact } inability to express herself ⊃ vary rhythmical pattern

[patient recognises rhythmical pattern] ⊃ { valid: plays dialogue } able to express herself communicatively ⊃ maintain dialogue

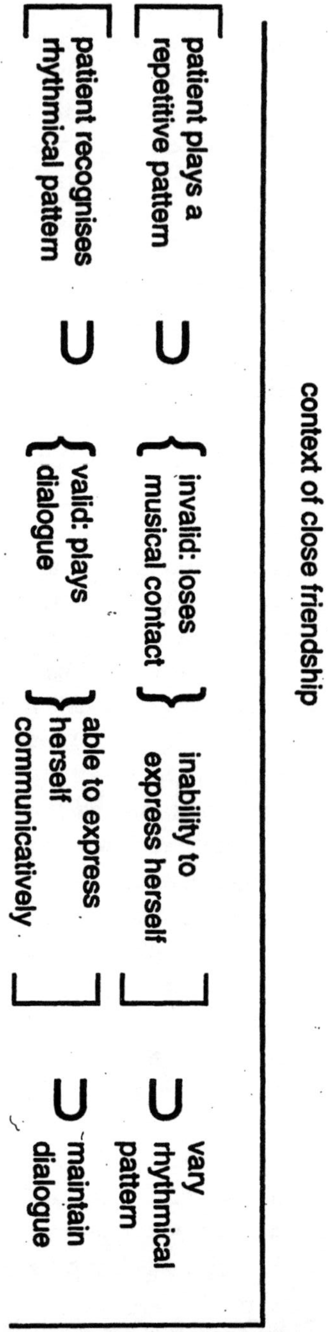

Figure 9.4 Regulative rules relating to the rhythmic improvised playing of Alzheimer's patients and its meaning for dialogue in therapy

CASE ELEVEN

Preverbal Communication Through Music To Overcome A Child's Language Disorder

AMELIA OLDFIELD, M.Phil., R.M.Th.
Music Therapist
Child Development Centre
Addenbrookes Hospital
Cambridge, Great Britain

Abstract. This case describes two years of group and individual music therapy for a five-year old boy with a language disorder. A wide variety of music therapy techniques are used, all aimed at motivating Jamie to communicate, either nonverbally or verbally.

BACKGROUND INFORMATION

Jamie is the only child of very caring and capable parents. As a young child, he appeared somewhat smaller and slower than other children in his age group, and eye contact was often difficult. His mother reports that, as a baby, he did not babble at all, and used very few other non verbal means of communication, such as pointing. He was always a very quiet child, and only occasionally and inconsistently used words.

At two and a half, Jamie found mixing with other children very difficult, and would often appear to be in a world of his own. However, he did not present any major behaviour problems and was able to play by himself. At this stage, the pediatrician reassured his parents that Jamie's development was not necessarily abnormal. Nevertheless, both his parents and other professionals involved continued to be concerned. Jamie's health visitor wrote a report at this time describing him as: "rather worrying in a not altogether definable way."

When Jamie was three, he was assessed by a clinical psychologist who suggested that, although his overall intelligence was within the normal range, there were great discrepancies in his skills. He had high scores for manipulative skills, such as putting puzzles together, and marked problems with both comprehension and expressive language. Jamie's hearing was also tested at this stage as he seemed both oversensitive to some sounds and oblivious to others. It was found to be within the normal range.

Jamie was then referred to the Child Development Centre where he began having regular sessions with the speech therapist. He also started attending a small play therapy group of four children. This is a structured group run by a clinical psychologist where the emphasis is on encouraging social integration. She reports that, over a period of a year, Jamie took part in more group activities and managed to overcome some of his fears and obsessions. He became more able to tolerate the screaming of another child in the group, for example, which he had been terrified of at first.

When Jamie was four, he was assessed by the local specialist consultant in child psychiatry. He suggested that, although Jamie's language was very restricted, he was showing signs of imagination. In spite of his difficulties, Jamie seemed to be developing an understanding of the meaning of words; therefore, there seemed to be potential for the development of abstract thought. The psychiatrist felt that Jamie's social problems and occasional disturbed behaviour were the result of his great difficulties in understanding social practices. Thus, he diagnosed Jamie as having a specific language disorder. In his opinion, there was no evidence of autism or an autistic like disorder.

The term "language disorder" generally describes an atypical pattern of language acquisition and development. Unlike children whose language may be delayed but nevertheless following a normal pattern of acquisition, children with language disorders have both a delayed and deviant pattern of development (Webster & McConnell, 1987). Deviancy or disruption may occur in any or all aspects of speech and language: context, form or use; or as a result of a distorted interaction between them (Bloom & Lahey, 1978). Jamie had difficulties in all these aspects of language development, and particularly in the area of language use. This affected his ability to establish social relationships and to relate to the world around him.

A speech therapy report written a couple of months later agreed with this diagnosis. The speech therapist explained that Jamie had difficulties processing sentences in order to comply with a task. Although he responded to everyday instructions, he was reacting more to the context and the routine than to the actual meaning of the instruction. Jamie could say quite long sentences, but had difficulties learning when to use these sentences appropriately. He was mainly silent and only made occasional, spontaneous, self-generated comments.

From the age of four to the present, Jamie has been attending a small language unit for eight children with language disorders. The children in this class receive very specialised schooling, and the main focus of the work is on improving their language difficulties. The class is based in an ordinary school, and the children are integrated into other "normal" classes at times, as well as working together as a group at other times. Both the teacher from the unit and Jamie's parents are still unsure about Jamie's diagnosis, and suspect that he might have some autistic tendencies.

At five, Jamie was referred to me at the Child Development Centre by his language unit teacher. She had noticed that Jamie seemed to respond to words in songs more easily than spoken words. She hoped that I might devise some exercises for both her and Jamie's parents to use with him to improve his speech.

MUSIC THERAPY ASSESSMENT

I saw Jamie for three consecutive weekly, half hour music therapy assessment sessions. The purpose of these sessions was: (1) to determine whether music therapy would be a useful way of helping Jamie, and if so to roughly outline what kind of direction this treatment might take; (2) to see whether he responded to me in a different way through music and thus to shed new light on some of his difficulties; and (3) to suggest ways in which both his teacher and his parents could use music with him.

Jamie presented as a small, attractive looking boy with a serious and often puzzled expression. He had no difficulties separating from his mother, and showed no anxiety about coming into the music therapy room with me. He seemed to understand simple requests or comments such as: "Here is a chair for you, Jamie" or "Shall we finish this now?" He was able to point to me and to choose an instrument for me on request. He could listen to my

playing and also play himself and was good at taking turns with me. He made very few verbal contributions or vocalisations, but at one point suddenly and surprisingly, made an appropriate comment about an instrument, saying in a very clear voice, "There's a ball inside."

Jamie particularly enjoyed activities where we teased one another, or where he could "control" me by, for example, making me jump when he played the drum. At these times, he would look straight at me and have a beautiful mischievous smile.

Jamie seemed pleased to listen to the music and the songs I improvised on the piano and the clarinet. He anticipated the ends of harmonic phrases by looking up at the appropriate moment, and showed that he knew and recognised a number of songs by occasionally filling in words when I left a gap. For example, I would sing: "London bridge is falling....," and Jamie would say: "Down!" Sometimes he would sing the words at the correct pitch to fit in with the song.

Jamie enjoyed playing the instruments, and would spontaneously explore various ways of playing them in a creative way. For example: he seemed to experiment with the different sounds the drumstick made on various parts of the drum, and played the cabasa in a number of ways, stroking and rattling the beads as well as shaking the whole instrument.

Jamie generally seemed to prefer the quieter instruments. He did not appear particularly frightened of loud sounds, but would blink slightly anxiously when they occurred. With a little encouragement, he could join in and enjoy both quiet and loud improvisations. He was able to follow dynamic changes when we improvised together, but had more difficulty following rhythmic changes. He appeared to be able to play in a regular pulse for short periods, but the pulse was hesitant and gave his playing a slightly tentative feeling.

Jamie found it difficult to move freely or spontaneously to music. His physical reactions were slow, and he needed encouragement to do things such as march or jump to the music.

Jamie seemed to be developing a positive relationship with me. He was at ease playing the musical instruments, and was able to both listen to and contribute musical ideas during our improvisations. I felt he would benefit from a situation where he could communicate with an adult without having either to understand spoken language or use words himself. The areas I thought we could work on were: increasing his motivation to communicate with another person; providing an opportunity for Jamie to vocalise freely and spontaneously; increasing Jamie's confidence and enabling him to speed up his reactions so they were more spontaneous. I therefore recommended that he should have weekly individual music therapy treatment for at least six months.

Jamie appeared to be more spontaneous in his communication with me during our sessions than he was with other adults. This was probably because far less speech was necessary in my sessions than in other situations. The fact that he was more at ease in this non verbal situation seemed to confirm the diagnosis of language disorder. After reading Jamie's notes, I had expected him to be more sensitive to loud sounds and was surprised when he did not seem to mind hitting the drum very loudly. On reflection, however, it became clear that it was unexpected and unexplained loud noises that particularly troubled Jamie, and not loud sounds that he knew were about to occur or sounds which he himself produced or controlled.

I did not think that it would be beneficial to give Jamie's parents or his teacher structured musical exercises to improve his speech. I felt the priority was to help Jamie feel at ease with a non verbal means of communication, so that he would eventually become more spontaneous in his efforts to communicate. I also thought that Jamie should be encouraged to enjoy making sounds and vocalising without the pressure of using the correct word or structure. Jamie had never babbled or experimented with sounds as a baby, and I thought that he needed to discover the fun of producing sounds. I therefore suggested that both his parents and his teacher should encourage Jamie to vocalise in any way, and that they should try to engage him in playful vocal dialogues. I also suggested they do "toddler" like rhymes such as "Incy Wincy Spider" or "Round and Round the Garden" with him, so that Jamie could laugh at them with an adult, and learn to enjoy communicating in a simple way.

TREATMENT PROCESS

Phase One: Introductory Group Work

Unfortunately, I did not have any spaces available to see Jamie for individual music therapy sessions immediately, and he was therefore put on a waiting list. As it happened, however, I had already arranged to see the group of children in the language unit that Jamie attended for a a twelve week period, starting four weeks after I had finished the assessment on Jamie. I was, therefore, able to observe and work with Jamie in a group setting before I started to work with him individually.

The group sessions occurred once a week, lasted approximately forty minutes and went on for one school term (twelve weeks). Both the teacher and the welfare assistant took part, and I reviewed our work with the teacher every week, directly after the session.

All eight children in the group were diagnosed as having language disorders. Jamie, however, was shyer and more withdrawn than the other children. The group sessions had two or three specific aims for each child. These were determined jointly by the teaching staff and myself after a couple of "exploratory" sessions. Generally, the goals were: to provide a different setting for the teaching staff to observe the children's strengths

and difficulties, and to give the teaching staff ideas of musical activities to use in the classroom. Given my large case load, this is one of the only ways I can provide some input to a large number of children.

The musical material and the activities used in the group would vary from week to week, and was largely determined by the aims for individual children. Although suggestions for activities for the following week's session might be made when we reviewed our sessions, I would always remain flexible and would usually choose activities on the spur of the moment, based on the children's reactions and moods on any particular day. Nevertheless, I would always start off with a familiar greeting song and end with a "good bye" activity. Throughout the group I would often alternate between activities which involved the group as a whole and activities which involved one or two children playing on their own. An example of a general group activity would be: the whole group plays together on various percussion instruments led by improvised music I play on the piano. When the piano stops the children all move around and exchange instruments. Playing starts again when the piano begins. An example of an activity involving two children would be: two children sit back to back in the middle of the circle, each with a different instrument, and are asked to have a musical conversation. The rest of the group is encouraged to listen. I would also try to alternate between activities where the children were actively involved in playing instruments, singing or dancing, and activities which required concentrated listening without so much active involvement.

After observing Jamie within the group for two sessions it became clear that he was much more withdrawn in this setting than he had been with me or a one to one basis. We therefore decided that individual aims for Jamie would be: to help him concentrate and listen to instructions; to encourage him to communicate in any way with either adults or children; and to encourage him to make eye contact and to make any vocal sounds.

During the first five sessions, Jamie seemed to understand some but by no means all the instructions, and was able to take part in a few activities only. He seemed to enjoy choosing and playing instruments, but was unable to pass an instrument to another child. He did not understand the games involving drama where we pretended to put a tambourine to sleep, for example, and he needed help whenever any of the activities involved moving around the room. He made little eye contact, and only used a few sporadic single words. He often appeared to be in a world of his own, and made no efforts to communicate with either the children or the adults in the group.

During the sixth session, there was a marked change in Jamie. He suddenly appeared more at ease, smiling happily and looking straight at me when I played the clarinet. He was able to contribute some vocal noises to a song where all the children were suggesting different sounds, and even gave his instrument to another child when this was suggested to him. From this session onwards, Jamie continued to progress well. He learned how to "conduct" by pointing to other children and adults. He would listen to instructions better, and he began to take part in even quite complicated activities. He started using more words, both on request and spontaneously. Both Jamie's teacher and I were pleased with Jamie's progress within the group, however we felt that he would benefit even more from individual sessions.

Phase Two: Individual Sessions

Two weeks after the group finished, a space became available, and I started to see Jamie for regular weekly individual music therapy sessions. Although he had made some progress during the group sessions, the aims remained the same: to increase communication, eye contact, vocalization, and spontaneity. As Jamie's use of words had improved, I continued to keep a record of both spontaneous speech and the speech he used to answer direct questions. Nevertheless, I still did not want Jamie to feel that this was the focus of our sessions, or that I was putting pressure on him to talk.

The individual sessions lasted half an hour, and were held at the same time and in the same room every week. After each session, I would briefly discuss with Jamie's mother how he was progressing.

Like the group sessions I would start and end each session with familiar "hello" and "good bye" activities. In between, sessions would vary from week to week depending on Jamie's mood, on what had happened the previous week, and in what particular areas I felt I should be helping Jamie. In general, I would spend some time encouraging him to choose instruments or activities, and then attempt to follow and support his playing; at other times, I would make suggestions myself. For example, I might suggest that we take turns playing the glockenspiel, and pass each other the stick when our turn was finished; or I might encourage Jamie to play three different instruments that would make me jump, wave my arms or shake my head depending on which instrument he played; or I might suggest that we have a "noise" dialogue on the kazoos. I would always try to give each of our activities a structure with a clear ending. I would prepare Jamie for each ending by saying "One more turn each," or "Try to find a way to finish this off."

Jamie was at ease with me straight away, and was delighted with the familiar "hello" song on the guitar. This led to a sung "noise" dialogue accompanied by shared guitar strumming. Jamie initiated vocal sounds such as "Hey" with great delight, and would then laugh happily. He gradually added "funny" faces to these noises, particularly when I encouraged him by mirroring and extending his contributions. Jamie was clearly excited and pleased with these humorous exchanges, and I was able to keep them immediately followed my greeting to him, and became a regular part of our sessions. Sometimes Jamie would respond immediately, and at other times it seemed to take him a little time to relax and allow himself to enjoy this basic form of communication.

Over the first six months of treatment, Jamie continued to become more spontaneous in any familiar activities that we shared. However, he would revert to a blank, puzzled expression whenever I introduced anything new. I, therefore, made a conscious effort not to allow the sessions to become too stereotyped and, while always keeping some familiarity, tried to vary the way we played together, always introducing at least one new idea every week.

As Jamie became more able to make his own choices and contributions, he started to use more single words or two word phrases, both spontaneously and in answer to direct questions. In a conducting game, Jamie gradually managed to give me more and more complicated instructions, such as "Play the drum and the cymbal loudly." Nevertheless, his speech was still far from normal, and at times he would be unable to say something as simple as "Goodbye, Amelia" or tell me which day of the week he came for music therapy.

Jamie still found it difficult to move quickly or spontaneously. However, he started to enjoy and understand imaginative games where I pretended to fall asleep on the piano, or I hid from him in the room. At these times he could react quite fast to "Wake me up!" or "Find me!"

Jamie continued to enjoy experimenting with various ways of playing the instruments, and seemed to become more sensitive to various tone colours. He began to listen much more carefully to the sounds he produced. His sense of rhythm also improved. He would enjoy improvising on the piano and quickly became able to pick out tunes such as "Ba-Ba Black Sheep," "Happy Birthday To You" and "Puff the Magic Dragon." As he apparently wanted to learn more tunes, and enjoyed playing the piano, I arranged for him to start piano lessons with a teacher who had an interest in children with special needs. This also meant that there could be a clear separation between my work and more formal piano teaching.

As Jamie gradually became more spontaneous in his contributions, he also developed some slightly obsessive behaviours, such as repeating a tune fragment again and again, or insisting on holding the drumstick in a certain way. Nevertheless, he could be distracted from these obsessions relatively easily. As time went on, these rituals seemed to die away, and were replaced by ordinary "toddler like" naughtiness and rebelliousness. The only slightly strange behaviour that did occasionally creep back was that of Jamie "telling off" his right hand for misbehaving.

By the end of six months of individual music therapy sessions, Jamie had made great progress, and the aims set out at the beginning of our work together had been achieved. Progress had also been noticed at school, and at home Jamie's parents were delighted with his greater willingness and ability to communicate. However, they were also finding him a great deal naughtier and less easy to manage.

I therefore decided that, as I had developed such a good rapport with Jamie, I would continue to see him for another four months with a view to helping both Jamie and his parents to cope with these new "naughty" behaviours. I also thought that his communication skills could be further improved.

Phase Three: A Slightly New Direction

Aims for the last four months of treatment were: to diminish silly behaviours such as screaming or deliberately throwing objects; to encourage longer spontaneous and creative dialogues with me (nonverbal and verbal); and to help Jamie to answer questions appropriately (and not let him divert me from this).

When dealing with Jamie's "naughty" behaviours, I felt it was important to explain what I thought about these behaviours, and why I was responding in a particular way. I told him that we would work out ways of stopping his naughty behaviours together. At times, I would smile at him, and tell him in a "teasing" way that I thought he was trying to be naughty. At other times, I would suggest to him that it was easier to opt out of an activity and be naughty, than to continue our work. When he threw an object, I would take his hand and physically help him to pick it up again, saying that it was important for us both to make the naughty behaviour "better."

Occasionally he would get "stuck" when asked to do something and say "I can't." In this case I would either help him physically (and comment that I was giving him a "helping hand"), or I would say that perhaps what I had asked Jamie to do was too difficult. This approach seemed to work well. He remained mischievous but became more accepting of direction, and would allow himself to be diverted from whatever was causing a problem more easily.

During the last few sessions, Jamie sometimes became "moody," and on one occasion, he cried when he did not have time to play an instrument he had wanted to play. He seemed relieved to be told that there was nothing wrong with being sad and crying.

During the last four months, Jamie continued to make progress in his communication skills. By the end of my time with him he was able to hold ordinary conversations with me. He would initiate a conversation and ask appropriate questions. However, he would still sometimes need encouragement to answer questions.

Overall, the progress he made during that year was remarkable. From a quiet often mouse-like child, he had become a vocal, boisterous child, often full of mischief and fun.

DISCUSSION AND CONCLUSIONS

In the first instance, the musical instruments and our music-making interested Jamie, and motivated him to be actively involved with me. This enabled me to start building up a relationship with him which was initially based on shared enjoyment of the music and the musical activities. Jamie was able to maintain this positive relationship with me because I used very little speech in our assessment sessions. He could, therefore, relax and simply enjoy being with me. We were playing music together and communicating through sound, but very few specific words needed to be said or understood. It was the use of music as a means of communication which was essential at this point, and this could only have been achieved through music therapy.

For the first few group sessions, Jamie again became very shy and withdrawn. This was probably because far more speech was necessary in this situation to understand what was going on and what was expected of him. However, he was able to maintain an interest in the group because of his fascination for music. The familiar structure of the sessions gradually reassured him, and gave him the confidence he needed to take part with the other children and make his own contributions.

When I started working with Jamie individually, the familiar framework of a "hello" and "good bye" activity reassured him, and allowed him to start work with me straight away. In fact, it became clear that Jamie relied too heavily on familiar and predictable activities, and I had to start introducing "surprises" so that he did not become entirely dependent on this familiarity.

One of the most important things that we worked on throughout Jamie's individual sessions was vocalisation. As Jamie had never babbled as a baby, I felt that he needed to discover what fun it could be producing sounds and experimenting with different vocal noises. It is interesting to note that it was during these vocal exchanges that Jamie first started using his face in an expressive way, wrinkling his nose and making "funny" faces. This ability to encourage a child to have vocal sung dialogues which can be varied and made interesting through musical improvisation is unique to the music therapist.

Another important aspect of our work was the fact that I was able to put Jamie "in control" by encouraging him, for example, to conduct my playing. I think this was helpful in building up Jamie's confidence, as his language difficulties often made him feel confused and "out of control."

Slowly, and almost in spite of himself, Jamie discovered that it was not only easy to communicate with an adult, but that it could be fun and therefore worth the effort. This was my main aim with Jamie but it happened so gradually that I only realised how much progress he had made when I looked back at how little he had initially contributed.

Finally, it is interesting to note that as Jamie's abilities to communicate improved, he developed new "naughty" behaviours. The approach that I used to help him with these behaviours was based on explaining my actions very carefully, and making use of his new found language and comprehension skills. At this stage, I was also able to put more pressure on Jamie and be more

Group Music Therapy
With a Classroom of 6-8 Year-Old Hyperactive-Learning Disabled Children

JULIE HIBBEN, M.ED., CMT-BC
Director: Programs for Special Needs
Powers Music School

Music Therapy Faculty
Lesley Graduate School
Cambridge, Massachusetts

Abstract: In this case, the author recounts the progress toward group cohesion of an early elementary special education classroom. Most of the children are described as Attention-deficit Hyperactivity Disordered. In the twice-weekly sessions during the year, the therapist uses active music making and movement to engage the children in interactive play, and to develop intimacy and cohesion in the group. The instruments, props, and songs serve as objects which encourage and contain the children's action and feelings. Developmental stage theory provides a framework for anticipating and planning group interactions, and for evaluating individual progress. Group activities are described in terms of five dimensions: interaction, leadership, movement, rules, and competency.

---

demanding, something I would have avoided doing in the earlier stages. I think it was my relationship with Jamie which was crucial at this point, rather than the special skills that I have as a music therapist. Nevertheless, I had developed this relationship through our music-making, so it was important for me to continue and complete our work together.

When I recently telephoned Jamie's family one evening to find out whether they would be happy for me to write this case study, I heard a familiar voice in the background: "I don't want to go to bed!" Certainly, this is a well-known and unwelcome communication for any parent to receive from a child, but in this instance, I could not help feeling moved. I was reminded of the amount of progress Jamie had made since I first saw him two years previously, when he had hardly been able to use speech to communicate in any way at all.

Although I generally enjoy my work as a music therapist, I do sometimes wonder whether I am really achieving results, and whether the children could equally well be helped through means such as special teaching or play therapy. Cases like Jamie make up for the times when progress seems to be very slow or nonexistent, and help to maintain my belief that music therapy is a truly unique and invaluable form of treatment.

REFERENCES

Bloom, L., & Lahey, M. (1978). <u>Language Development and Language Disorders</u>. New York: John Wiley and Sons.

Webster, A., & McConnell, C. (1987). <u>Children With Speech and Language Disorders</u>. London: Cassel.

BACKGROUND INFORMATION

The eight children in this self-contained classroom had chronological ages of 6-8 but in most cases they were academically at pre-first grade level. Many of the children had disruptive behavior disorders associated with *Attention-Deficit Hyperactivity Disorder (ADHD)* as well as *Learning Disabilities*. In some cases the children were at risk because of the severity of their anti-social behaviors and/or the lack of support from their family systems. Some of the children were on psychostimulant drugs such as Ritalin which controlled their hyperactivity to some extent. The children lacked developmental experiences such as nurturing play and spontaneous game playing with peers, either because of environmental deprivation or because of their learning or behavioral disorders. Their behaviors ran the gamut from excessive activity, interruptive talking, and physical aggression to negativism, lethargy, and introversion.

The brief description of the children below points up the diversity of their needs and behaviors:

Paul came from a special needs preschool, and at age 6.5, had just entered the classroom. He had no reading skills and had expressive and receptive language disabilities so severe that *auditory aphasia* was being considered as a diagnosis. Paul learned visually. Perceptual motor disabilities were apparent in his awkward maneuvers around the classroom. His relationship to his father was close and active.

Arnie had a history of depression and low self-esteem. His interaction during the music therapy assessment showed him to have ability (fluency, memory, originality) and confidence in expressing himself musically.

Al was moody and sulked a lot. He had no reading skills and had expressive and felt threatened by others in the group, even when the threat was not warranted. His voice was often raised in complaint.

Nathaniel had a history of passive aggressive behavior. He used his intelligence and verbal skills to manipulate those around him. Nathaniel, 8 years old, left the class after the first month.

Ken was dependent and passive, encouraged in this by his doting parents. He was medicated with Phenobarbital which slowed him down motorically. His thinking was very concrete. Ken was not able to verbalize his feelings, which often led to violent outbursts. His body was overweight and flaccid, and he showed signs of perceptual-motor disabilities.

Michael was anxious, hyperactive, and intelligent. During the middle of the year his cousin was taken to court for continually sexually abusing him.

Jose was humorous and friendly at times, and at other times was negative and depressed. During initial music therapy assessment, Jose was in control of the interaction, though he interacted only with expressive mime and eye contact. Jose left the classroom at mid-year.

Hattie (the only girl) was 6 years old. She was defiant, and deliberately

CHAPTER 5

Individuals with Autism and Autism Spectrum Disorders (ASD)

Mary S. Adamek
Michael H. Thaut
Amelia Greenwald Furman

CHAPTER OUTLINE

DEFINITION AND DIAGNOSIS
ETIOLOGY
CHARACTERISTICS
    Communication
    Social Interactions
    Sensory Processing
    Behavioral Issues
MUSIC THERAPY GOALS AND INTERVENTIONS
    Music Therapy to Improve Communication Skills
    Music Therapy to Improve Social and Emotional Skills
    Music Therapy to Improve Behavior
    Music Therapy to Improve Academic, Physical/Motor, and Leisure Skills

Autism is a **pervasive developmental disability** that affects a person's ability to communicate and interact with others. The term *autism spectrum disorder* (ASD) refers to the fact that this condition affects individuals in different ways and to varying degrees (Autism Society of America, 2008). ASD affects approximately 1 in 150 people in the United States (Rice, 2007), which means that approximately 1.5 million Americans today have some form of autism. It is four times more prevalent in boys than in girls, and the rate of children diagnosed with autism is increasing by 10–17% each year. At this rate, the number of people with autism could reach 4 million within the next decade (Autism Society of America, 2008).

Many music therapists work with children and adults with autism in schools, residential settings, and community agencies. According to the U.S. Department of

117

Education (2007), over 92,000 students ages 6–17 with autism received some form of special services under IDEA during the 2001–2002 school year. (See Chapter 14 for more information about IDEA.) The majority of music therapy services with individuals with autism are provided to children, primarily in school and community settings. Adults with severe autism may receive music therapy services in residential facilities for people with developmental disabilities, or they may receive services in community settings with other adults who have developmental disabilities. This chapter will focus on the characteristics and needs of children with autism and music therapy interventions for children with autism. For more information on music therapy with adults who have developmental disabilities, refer to Chapter 4.

DEFINITION AND DIAGNOSIS

Leo Kanner (1943), a psychiatrist at Johns Hopkins University, was the first person to identify autism as a distinct developmental disorder. He described a group of children who were relatively normal in physical appearance but who exhibited severely disturbed behavior patterns that included the following: extreme social aloofness or aloneness; lack of emotional responsiveness; avoidance of eye contact; failure to respond to auditory or visual stimulation; lack of language development or failure to use language adequately for communication; excessive attachment to objects; and preoccupation with ritualistic, repetitive, and obsessive behaviors. Because the symptoms were presented in early infancy, Kanner coined the term *infantile autism*.

Since those early years, the definition and diagnostic criteria of autism have been changed and refined, mostly in terms of broadening the definition. Autism is described as a spectrum disorder, meaning that there are a variety of possible disorders and characteristics having differing levels of developmental delay ranging from mild to severe. Thus, generalizations are difficult to make since individuals with autism can be very different from one another, and there is a wide range of abilities and deficits.

Autism Spectrum Disorder (ASD) falls under the umbrella category of **Pervasive Developmental Disorders (PDD)**, listed in the *Diagnostic and Statistical Manual of Mental Disorders* (American Psychiatric Association, 2000). This category includes neurological disorders such as autism, **Rett's Syndrome,** and **Asperger's Syndrome**, which are characterized by severe and pervasive impairments in many areas of development (American Psychiatric Association, 2000; Autism Society of America, 2008). Autism and ASD affect children of all social classes, financial levels, educational levels, cultures, and races throughout the world (Autism Society of America, 2008; Scott, Clark, & Brady, 2000).

CHAPTER 5: INDIVIDUALS WITH AUTISM AND AUTISM SPECTRUM DISORDERS (ASD) — 119

Persons with autism have serious deficits in communication and social skills, and they may display behaviors that are unusual compared to their typically developing peers. Symptoms of autism begin before the age of 3 and continue throughout a person's life. All of the following diagnostic criteria must be present from early childhood in order to make a diagnosis of autism:

- *Qualitative impairment in reciprocal social interaction,* based on what is typical for developmental level. This may manifest in poor eye gaze, disinterest in personal relationships, and limited use of gestures.

- *Qualitative impairment in verbal and nonverbal communication,* based on what is typical for developmental level. This may manifest in language acquisition delay, limited or lack of speech, lack of spontaneous and varied make-believe play.

- *Restricted repertoire of interests and activities,* based on developmental level. This may manifest in stereotypical or repetitive movements such as rocking, hand flapping, or spinning as well as limited or abnormally intense interest areas. (Frith, 2003)

As a spectrum disorder, the range of abilities and degree of developmental delay result in unique profiles among persons with ASD; however, all children with autism have some sort of difficulty with communication, social skills, and behavior (Frith, 2003; Johnson, 2004; Mastropieri & Scruggs, 2000; Scott et al., 2000). Those who have difficulty with verbal communication and expressing needs may point, use gestures, or use other nonverbal forms of communication such as pictures or icons. Many, but not all, children with ASD also have cognitive impairments; however, children with Asperger's Syndrome have high cognitive abilities combined with severe impairment in reciprocal social interaction and restricted interests. A person with autism may have a combination of the characteristics listed in Table 1 on the following page (Centers for Disease Control and Prevention, 2008), which will be discussed more extensively later in this chapter.

Table 1
Continuum of Abilities and Limitations for Persons with Autism

AREA OF FUNCTIONING	VARIABILITY LEVELS
Measured Intelligence	Severely impaired ----------------Gifted
Social Interaction	Aloof ------Passive -----Active but odd
Communication	Nonverbal -------------------------Verbal
Behaviors	Intense ---------------------------------Mild
Sensory	Hyposensitive ----------Hypersensitive
Motor	Uncoordinated -------------Coordinated

Individuals with autism or ASD may display some of the following specific traits or characteristics. It is important to note that not all people with this diagnosis have all of these traits, and the severity of the disorder varies by individual (Autism Society of America, 2008).

- Repeating words or phrases, **echolalic** language
- Unresponsive to verbal cues or directions; may appear to be deaf due to unresponsiveness
- Difficulty interacting with peers; minimal spontaneous socialization
- Oversensitivity or undersensitivity to stimuli or to pain
- Resistance to change; insistence on routine
- Minimal direct eye contact
- Odd or unusual play, particularly sustained play or attachment to objects

Contrary to common beliefs, children with ASD may make eye contact, develop good functional language and communication skills, socialize with peers, and show affection. Their skills in these areas might be different or less sophisticated than typically developing peers, but it is very likely that with appropriate educational and social interventions, children with ASD may develop functional and appropriate skills in all of these areas. Other behavioral or emotional disorders may co-occur with ASD as a secondary diagnosis, including obsessive/compulsive disorder, anxiety disorder, depression, and attention-deficit/hyperactivity disorder. (See Chapter 8 for a review of behavioral-emotional disorders.)

ETIOLOGY

When autism was first recognized as a distinct disorder, psychiatrists attributed it to an early emotional trauma or to faulty parenting. The unusual array of symptoms (i.e., relatively normal physical appearance and isolated skills coupled with the emotional unresponsiveness, social aloofness, and difficulty with language) led many researchers to believe that autism was the result of an emotional trauma in very early childhood. However, because no consistent patterns of social or emotional history emerged in these children, other etiological considerations emerged.

Since the 1960s, accumulated research evidence strongly suggests that autism is a developmental disorder of brain function that is manifested in a variety of perceptual, cognitive, and motor disturbances. To date, there is no single known cause of autism; however, there is general agreement that autism is related to abnormalities in the brain structure and function. Researchers have found differences in brain scans between brains of individuals with autism and individuals with typical development (Frith, 2003). Research is underway to test several theories related to the causes of autism. Some theories link the disorder to genetics and heredity, while others cite environmental risks as causal factors related to the development of autism. Currently, many national organizations are funding and conducting research to determine the causes of autism and best practices for intervention (Autism Society of America, 2008; Autism Speaks, 2008; Centers for Disease Control and Prevention, 2008).

The following sections describe three functional domains in which children with autism often have difficulties. Typical educational approaches to support learning will be described.

CHARACTERISTICS

Communication

As noted previously, a child with autism has qualitative impairment in language and communication skills. Some children may have no ability or interest in communicating verbally or nonverbally with others. Some have **echolalic** language, in which previously heard words or phrases are repeated without any intent to convey meaning. Children with autism have a marked difference in language skills compared to children who are typically developing. Impairments may be characterized by:

- lack of spontaneous social imitation (e.g., waving and saying "goodbye")
- failure to use language correctly (e.g., syntactic problems, not using verbs)
- limited vocabulary and semantic concepts, poor intonation patterns
- pronoun reversal (e.g., saying "you" instead of "I")
- lack of gesture or mime when trying to make needs known
- difficulty understanding spoken language

- failure to develop joint attention skills in preverbal children (e.g., pointing to a toy to share one's pleasure with another person). (Heflin & Alaimo, 2007; Mundy & Stella, 2000)

Children who have severe levels of autism may have little or no receptive or expressive language, while students with mild levels of autism may have developed language skills that allow them to communicate with others. Because communication is a primary deficit for individuals with ASD, the development of communication skills is often a key focus for therapy. This may include the development of verbal communication or effective use of alternative or augmentative communication systems.

There are many **alternative and augmentative communication (AAC) systems** available, which may enhance a child's verbal communication skills or become a child's primary means of communication. These alternative systems help a child express wants and needs, initiate and maintain conversations, and receive and understand information from others. AAC systems can be simple systems of pointing to pictures, words, or letter boards, or more sophisticated methods, such as use of sign language, voice output computers, or visual tracking devices (National Research Council, 2001; Scott et al., 2000). These communication systems can provide a means for students to communicate with others in a meaningful way and can facilitate two-way, interactive communication between the student with autism and others.

The **Picture Exchange Communication System (PECS)** is a good example of an AAC system that is used in many homes and classrooms. PECS is system that teaches children to use pictures and symbols in order to ask for wants and needs, respond to others, and initiate conversations. Using PECS, a child learns the meaning of a set of pictures or symbols by exchanging those symbols for something he or she wants or needs. Using a sentence strip that says "I want _____," the child learns to fill in the blank with a picture (attached to the sentence strip with Velcro) that stands for whatever he or she wants; then the child hands the sentence strip back to the teacher, therapist, or other conversational partner in exchange for the desired object (Bondy & Frost, 1994; Frost & Bondy, 1994; Kravits, Kamps, Kemmerer, & Potucek, 2002). Other AAC systems might be a simple picture/symbol book, board, or wallet used for making requests or responding to questions. Some children may use basic signs to communicate with others. More sophisticated technology such as computers with voice output systems can also provide an effective means of communication. Music therapy implications for the use of ACCs will be discussed later in this chapter.

Social Interactions

Social skills play a major role in child development and support a child's ability to function in integrated classrooms or community settings. Appropriate social skills are necessary for successful interactions with peers without disabilities and for participation in normalized activities. Individuals with ASD have core deficits in social interaction and social perception, which manifest in difficulty initiating interaction, difficulty maintaining social relationships, and difficulty understanding the perceptions of others (Scott et al., 2000). Social deficits can range from mild (such as difficulty making and maintaining eye contact), to severe (such as the inability to share experiences and interests with others). Some individuals with less severe deficits may interact spontaneously with others, while those with severe social deficits may seem oblivious to others and the environment. Even individuals with higher functioning skills may have problems understanding the perspective of others, and understanding that the thoughts, beliefs, and intentions of others may be different from their own. Some children have difficulty with physical touch or in maintaining socially acceptable personal space. For example, some children stiffen and avoid physical contact, while others may be clingy and have inappropriate boundaries (such as always wanting to sit on the therapist's lap or in uncomfortably close proximity to others). Social behaviors can also vary over time.

Because social deficits are a key feature of autism, these children need direct training to improve social skills in a variety of settings. While children with autism can develop more socially appropriate behaviors through social skills training, it is important to note that communication skills provide the foundation for developing and maintaining social relationships; consequently, children with autism need to develop functional communication *along with* social skills in order to enhance interactive skills, make friends, and be included in classroom and social activities.

A variety of techniques can be used to teach social skills to children with these deficits. When choosing social skill interventions, teachers and therapists must keep in mind the age, language skills, developmental level, and interests of the child in order to develop effective approaches. Some current educational approaches used to teach social skills include:

- **Direct instruction**—teaching children directly to interact with peers with the support of prompts, modeling, or physical assistance during the interaction
- **Social communication training**—teaching children to ask questions and seek out information in a social setting
- **Use of social stories**—stories developed to teach social skill concepts and strategies (Gray & Garand, 1993); these stories can also be set to music to create social story songs

- **Leisure-related social skill development**—teaching social interaction and rules of play through leisure skills. (Scott et al., 2000)

Sensory Processing

Children who are typically developing have intact sensory systems that help them perceive the world around them. Their senses of vision, hearing, touch, taste, and smell work in an integrated fashion to help them make sense of their environment. Many children with autism have sensory processing problems. To a child with sensory processing problems, the environment may be confusing, painful, or even frightening. Some children with autism have sensory processing problems that can cause oversensitivity or undersensitivity to certain stimuli. For instance, some children with autism are particularly sensitive to high or loud sounds; when presented with this type of sound, they will shriek, hold their ears, or become aggressive in some way. Because they may process and respond to information in different ways, children with autism may exhibit atypical, inappropriate, unusual, or **stereotypical behaviors.** They may exhibit aggressive or self-injurious behaviors when they are unable to understand or unable to communicate their confusion. Keep in mind, however, that not all children with autism exhibit these types of behaviors.

Because musical sounds are such an integral part of music therapy, the processing of and response to musical sounds by children with autism is of particular interest. A number of research studies with this population document unusual sensitivity and attention to music. In fact, in 1964, a prominent researcher, Bernard Rimland, even listed unusual musical capabilities as a diagnostic criterion for autism (Rimland, 1964). Sherwin (1953), in a case study of boys with autism, noted strong melodic memory; recognition of classical music selections; and strong interest in playing the piano, singing, and listening to music. Observations of 12 children with autism over a two-year time period indicated heightened response and interest in musical sounds as compared with other environmental stimuli (Pronovost, 1961). Studies by Frith (1972), O'Connell (1974), Blackstock (1978), and Thaut (1987) showed improved task performance or attention on tasks involving music compared with those using other modalities.

Studies by Applebaum, Egel, Koegel, and Imhoff (1979) and Thaut (1987) indicated that children with autism could perform similarly to age-matched children with typical development on musical tasks involving imitation or improvisation. Koegel, Rincover, and Egel (1982) described music as an efficient motivator and modality for enabling autistic children to learn nonmusical material and emphasized its use as positive sensory reinforcement in decreasing self-stimulation behaviors.

Recent studies provide evidence that a high level of musical responsiveness can co-exist with low functioning in other cognitive areas in autistic children (e.g., Heaton, 2003; Heaton, Hermelin, & Pring, 1998; Heaton, Pring, & Hermelin, 2001; Hermelin, O'Conner, & Lee, 1997; Mottron, Peretz, & Menard, 2000). In addition, research examining the unique auditory perception of children with autism indicates higher performance in perceptual tasks involving music compared with other acoustic stimuli (e.g., verbal language [Foxton et al., 2003; Heaton, 2003, 2005; Heaton & Wallace, 2004; Mottron, Peretz, Belleville, & Rouleau, 1999; Nieto del Rincon, 2008; Young & Nettelbeck, 1995]). It is important to note, however, that some children with autism may show unusual and negative sensitivity to some musical sounds, or some combinations of music and movement.

It is, as of yet, unclear why many children with autism show heightened interest in and more typical responses to music than to other types of stimuli. However, the fact that music can be engaging and can support optimal functioning in children with autism explains in part why music therapy can be an excellent therapeutic choice for this population.

Behavioral Issues

Some children with autism have behaviors that are difficult to manage in a variety of settings such as the classroom, the home, or the community. Maladaptive behaviors include poor attention, aggression, stereotypic or self-stimulating behaviors, oversensitivity to sensory input, and difficulty with generalization of skills (National Research Council, 2001; Simpson & Miles, 1998). These behaviors can be among the most challenging and stressful factors for professionals and can interfere with learning and development in communication and socialization. Problem behaviors may be triggered or exacerbated by an inability to understand the expectations of the environment (including consequences of their behavior), inability to communicate wants and needs, or difficulty in initiating and maintaining positive social relationships. Aggression towards others, self-injurious behaviors, noncompliance, and disruption of session routines may create difficult and frustrating situations for everyone involved. Table 2 provides examples of the characteristic features of autism and how they might cause difficult behaviors (Adamek & Darrow, 2005, pp. 190–191).

Table 2
Characteristic Features and Examples of Difficult Behaviors

Characteristic Feature	Specific Difficulties	Possible Behavior
Language and communication impairment	Difficulty understanding directions and expressing self	Not following through with directions, aggression to others or self
Social interaction difficulty	Difficulty reading social cues from others, difficulty understanding the perspective of others	Difficulty interacting with others, sharing classroom materials, inflexible in social situations; isolation from peers, minimal spontaneous interaction with peers
Focus of attention problems	Difficulty processing information, focusing on the main feature of instruction; difficulty filtering out auditory/visual distracters	Not following through with directions, tasks, or assignments; engaging in off-task or self-stimulating behaviors; inability to follow multi-step directions
Aggressive, stereotypic or self-stimulating behaviors	Difficulty communicating needs or understanding expectations; frustrations	Acting out towards others; rocking or hand flapping to provide sensory stimulation; visually focusing on lights, fan blades, or shiny objects
Oversensitivity to sensory input	Processing problems that impair tolerance for tactile, visual, or auditory stimuli	Acting out or isolation; holding ears with hands; rejecting touch from others; refusal to play instruments that provide tactile stimulation
Difficulty generalizing behaviors	Difficulty applying skills learned in one situation to another situation	Difficulty with transitions; may act out physically, vocally, or refuse participation; difficulty transferring skills from one setting to another without additional instruction

Fortunately, problematic behaviors can be reduced by setting up a structured environment and through the use of behavioral strategies (providing examples and cues to clarify expectations, reinforcing positive behavior, etc.). Environmental structure (such as a well organized room with consistent work spaces and limited clutter) can help the child organize and focus attention appropriately. Some children derive structure and support by being close to others for structure and support, while others may need more space due to issues such as tactile defensiveness or hyperactivity. Providing instruction in close proximity is important for some children. Close proximity makes it possible for the music therapist to provide physical and gestural prompts when needed and gives the child a clear idea of where he or she should be looking for instruction.

In addition to setting up a structured environment, it is important to understand causes of inappropriate behavior so that strategies for developing appropriate behavior can be devised. For example, some children are more likely to misbehave if they sit next to one another—they either irritate one another or egg each other on. The mere fact of sitting by one another can trigger negative behaviors. A seating chart that places these children away from each other can prevent problems from beginning. For some children, sitting near the therapist may foster calm and focused behavior. It is important to recognize antecedents (what happens before the target behavior) and consequences (what happens as a result of the behavior) in order to determine some possible causes and reinforcing events.

While no single intervention will deal effectively with all problem behaviors, most professionals recommend using a preventative approach to decrease problem behaviors and increase positive behaviors. A proactive approach, such as utilizing **positive behavioral supports** (assessing what maintains the behavior and developing strategies to replace the behavior), can create an environment for success. This approach builds on the strengths of the child and focuses on changing the environment, instructional strategies, and consequences to promote positive behaviors.

MUSIC THERAPY GOALS AND INTERVENTIONS

Music therapy is an effective approach for addressing language and communication skills, social skills, cognitive skills, and behavioral skills for children with autism (Bettison, 1996; Davis, 1990; Edgerton, 1994; Goldstein, 1964; Hairston, 1990; Hermelin, O'Connor, & Lee, 1989; Hollander & Juhrs, 1974; Humpal, 1990; Kostka, 1993; Lim, 2007; Litchman, 1976; Ma, Nagler, & Lee, 2001; Mahlberg, 1973; Nelson, Anderson, & Gonzales, 1984; Reitman, 2006; Saperston, 1973; Schmidt & Edwards, 1976; Staum & Flowers, 1984; Stevens & Clark, 1969; Thaut, 1984, 1988; Umbarger, 2007; Walworth, 2007; Warwick 1995; Whipple, 2004; Wimpory, Chadwick, & Nash, 1995). Music therapy goals for children with autism focus primarily on improving communication, social interactions, and behavior. Secondary areas of focus include improvement of academic skills, physical skills, and leisure skills.

As noted earlier, some children with autism are more attentive and responsive to musical stimuli; thus, music therapy can be a highly motivating medium for addressing these goals. In addition, music is a flexible medium (see Chapter 3); therefore, it can be used for a variety of goals and readily adapted to suit the diverse strengths and limitations found within this population. Music therapy interventions involve the child in listening, singing, playing, moving, or responding verbally and nonverbally. Through goal-directed music therapy interventions, a child can also work on several skills at once. For instance, an intervention that involved participation in a percussion ensemble could focus on communication (nonverbal communication and self-expression through drumming, making choices about what instrument to play), social skills (interacting appropriately with peers, taking turns playing, listening to others, taking leadership) and appropriate behavior (following directions, listening to the leader, following through with tasks).

Music Therapy to Improve Communication Skills

As noted previously, children with autism have deficits in expressive and receptive communication skills. Music therapy provides a rich opportunity for language experiences through age-appropriate and interesting music interventions. Several studies indicate that music therapy can improve communication and language skills (Edgerton, 1994; Litchman, 1976; Mahlberg, 1973; Saperston, 1973). Music therapists use the elements of music (melody, rhythm, pitch, dynamics, form) to develop basic listening skills such as **auditory awareness** (sound vs. no sound), **auditory discrimination** (are two sounds same or different?), **sound identification** (recognition of the sound, such as a flute or a piano), and **localization** (locating the source of the sound). Music therapy interventions can encourage **expressive** (speaking, signing) and **receptive** (listening, understanding signs and gestures) **language**.

Because music is interesting and motivating, it can promote attention, active participation, and verbal and nonverbal response. For instance, action songs, chants, and instrument playing can elicit vocal/verbal and physical responses that promote receptive communication skills. Rhythm paired with speech can encourage verbalizations and appropriate pacing. Call-and-response rhythmic chants can increase imitation skills, first on instruments, then in speech. Songs with repetitive lyrics and melodies can facilitate memory for vocabulary and other information. Pairing songs with visual cues and movement can enhance comprehension of vocabulary and concepts. For instance, a song about winter could be used to introduce and reinforce concepts such as temperature, seasonal changes, appropriate attire (winter coats and mittens), and directions. Music can be a highly motivating tool to encourage imitative and spontaneous language (see Figure 1).

> **EXAMPLES OF MUSIC THERAPY INTERVENTIONS FOR COMMUNICATION**
> (Thaut, 1999)
>
> 1. **Music interaction to establish communicative intent (facilitate desire or necessity to communicate)**
>   - Offer musical interactions (e.g., question-and-answer, imitative) on drum, metallophone, or other instruments.
>   - Accompany the child's movements or habitual sounds (crying, laughing) on the piano.
>   - Sing an action song to the child and cue the proper physical responses.
>
> 2. **Action songs to promote interaction (after communicative intent is established)**
>   - Introduce chants or songs that integrate rhythm, body percussion, and vocalization that present instructions for active physical and vocal response.
>
> 3. **Oral motor exercises to strengthen awareness and functional use of lips, tongue, jaws, teeth**
>   - Play wind instruments.
>   - Performing oral motor imitation exercises of the articulators.
>
> 4. **Sequence imitation of gross motor, oral motor, and oral vocal motor skills (after perceptual and imitative skills are established)**
>   - First, introduce the name of a body part (e.g., "arm") while moving that part.
>   - Second, have the child practice oral motor positions of the articulators for that word (e.g., "a" and "m").
>   - Third, while the child moves the body part in an imitation exercise, ask him to sound out the parts of that body part (e.g., saying "arm" while moving, or parts of the word as is possible).
>
> 5. **Shaping vocal inflection of children who have some speech sounds**
>   - Vocal improvisation on vowel and consonant combinations.
>   - Sustaining sounds on wind instruments.
>   - Producing vocal sounds combined with graphic notations to represent speech inflection.
>   - Breathing exercises to improve vocal strength, exercise laryngeal function, and refine oral motor function.

Figure 1. Examples of Music Therapy Interventions for Communication Skills

If a child is using an **alternative and augmentative communication (AAC) system**, the music therapist can collaborate with the speech language pathologist to create responses for use in music. Examples include:

- Icons to represent music, or a picture of the music room
- Pictures or names of teachers, therapists, or peer buddies in music
- Pictures or drawings of instruments or props used in music
- Names of favorite songs sung in music

- Iconic representations of note values, such as whole notes, half notes, quarter notes, eighth notes
- Pictures related to the schedule or order of events in music, such as a "hello song" or opening song," "movement time," "play instruments," or "sing songs"
- Functional signs used for directions such as "stop," "play," "dance"

Because communicative deficits are a key problem area for children with autism, it is important to promote receptive and expressive communication in all aspects of instruction. Many children are more successful when presented with visual cues along with verbal instructions. Visual cues (**icons, photos, functional signs**) can be used for giving directions, offering choices, teaching new skills, or providing structure for an activity. Visual enhancements should be simple and consistent throughout the child's day to encourage generalization of skills and to support comprehension (Scott et al., 2000). Suggested uses for visual cues include:

- Describing rules (listen, hands to self)
- Presenting schedule of interventions, which creates predictability for children
- Modeling movement
- Pairing pictures with songs to allow choice (a picture of a sun for "Mr. Sun")
- Presenting a calming routine for coping with change or anxiety (e.g., take two breaths, squeezing/releasing hands). (Adamek & Darrow, 2005)

Abdi faces several challenges in his education. Not only does he have autism, but his native language spoken at home is not English. Abdi frequently becomes upset and aggressive in the classroom because he is unable to communicate his choices. Music therapy provides a consistent structure and the opportunity for repeated practice of words he needs to communicate choices. A calming song, written by his music therapist, lists six of Abdi's favorite things to do (e.g., rest, read a book). His classroom teacher keeps a copy of Abdi's special song (along with visual cues) in the classroom for use during choice time, or to help calm him when he becomes upset. The songs learned in music therapy are integrated into other instructional times to help hold Abdi's attention, increase his comprehension, and increase his time on task in the group. His ability to communicate choices about songs, instruments, and feelings has also contributed to his improved behaviors in school.

Music Therapy to Improve Social and Emotional Skills

Evan, a student with Asperger's Syndrome, entered the school as a 2nd grader. Although he was academically above grade level, he was having difficulty during lunch and recess. Some students found his loud voice and unusually deliberate speech patterns irritating and his social skills awkward. However, Evan was able to shine in a special classroom project, a musical written and produced by the students, about how blood circulates in the human body. A number of students auditioned for the "scientist narrator" position, but most could not be easily understood through the microphone and speakers. Evan, an excellent and precise reader, was perfect for the part. As a lead part in the musical, he was able to utilize his strengths and be seen as an essential member of the group, especially by the students writing the script. The friendships and tolerance that developed during the musical carried over to the creation of a small music group in which social skills training became a part of Friday Fun Time. Evan realized that during music, he understood the rules and could be successful with peers. He later joined band and excelled in that structured experience right up through high school graduation.

Children with autism and ASD have deficits in social skills, social interaction, and emotional expression. Because autistic children often reject or ignore others' attempts at social interaction, music can function as an attractive mediating object. Music provides a point of mutual interaction between therapist and child. Studies by Goldstein (1964), Stevens and Clark (1969), Hollander and Juhrs (1974), Schmidt and Edwards (1976), and Warwick (1995) have shown improved social behavior and interpersonal relationships as a result of music therapy treatment. In addition, music is an effective medium for eliciting and shaping emotional responses. Different types of music can be associated with different moods and emotions, which can in turn be paired with body language and verbal labeling of moods (Thaut, 1999).

Music making (singing, movement, instrument playing) with the therapist or with peers lends itself to interaction and the use of social skills. In one-to-one sessions with the child, the therapist can make musical contact and set up musical interactions that require social interaction (such as imitation or call and response) (Thaut, 1999). In group music experiences, children can practice responding to others, taking turns, listening, sharing ideas, greeting others, and sharing equipment. Simple skills such as holding hands with a peer, listening to others, or starting and stopping with the group can be difficult but important skills to develop. Below is an example of a social story song about *waiting your turn* (to the tune of ABC song) (Adamek & Darrow, 2005, p. 187):

Wait your turn, wait today, Then you'll have a chance to play.
First it's her, then it's him, Round the circle back to you.
Wait your turn, wait today, Then you'll have a chance to play.

Music therapists can promote meaningful social relationships with peers by providing structured and motivating opportunities for social interaction in the music setting (Adamek & Darrow, 2005; Brown et al., 1979; Jellison, Brooks, & Huck, 1984). Music groups including movement can also provide opportunities to practice appropriate personal space (how close one stands or sits next to others while still feeling comfortable), and socially appropriate touch.

Jamilia, who has autism, is a very musical child. She is able to learn new lyrics and melodies quickly and follows directions sung in lyrics well. Jamilia really wanted to be a part of the group, but her initiations were often rebuffed by other children because she had no sense of space or boundaries. She would squeeze herself into a classroom line, typically next to students wearing the prettiest dresses. Sometimes, out of the blue, she would lean over and touch or sniff other children's hair and talk about what they were wearing or doing. In music therapy, Jamilia learned songs that taught interactive skills, such as "find a friend and look them in the eye, find a friend and now say Hi."

For movement activities, the teacher marked the classroom floor with round Velcro pieces that were called "bubbles." These bubbles provided an easy visual reminder of where to stand in relation to others. In an intervention similar to the game Musical Chairs, the children would move as long as the music was playing. Whenever the music stopped, students jumped on to one of the bubbles. Students were praised for "good bubble space." As the children internalized this sense of appropriate social space, the Velcro spots were gradually removed so students needed to "think the space." The concept of bubble space (socially appropriate personal distance) was then integrated into other favorite activities, such as singing a favorite train song, which required the children to keep their social distance while moving. These skills were then successfully generalized to other classroom and social settings.

Music Therapy to Improve Behavior

Abdi, like many children with autism, does not like change. Sometimes, he has verbal outbursts or becomes physically aggressive when his teachers change the classroom routine. Fortunately, his interest in music and music activities helps him to tolerate change. For example, Abdi has become accustomed to and enjoys playing drums with mallets; but because the music is so motivating, he is willing to try a different type of drum played only by his hands as requested.

Behavioral difficulties are another key feature of ASD. Just as with all other characteristics of autism, behavior skills vary in severity from one individual to the next. As noted earlier, behavior problems may be related to difficulty communicating or limited comprehension of the expectations and consequences of their behavior.

Music therapy can be structured to provide predictable experiences in which children can practice appropriate behavior. The rules for appropriate participation in a music group (such as following directions, starting and stopping at the correct place) can be practiced step by step, even as the group members engage in attractive musical arrangements. Those students with more advanced musical skills can practice these basic behavioral expectations within the context of more complex and interesting music. For instance, a small group percussion ensemble can learn basic steady beat patterns that can be extended to include more complicated rhythms on a variety of instruments. Children who follow directions, meet expectations, and participate with others may be asked to perform for others. In this way, music can be used to improve self-esteem and leadership skills. Music can also be used as reinforcement for appropriate behavior, offered as a reward for following through with expectations. Extra time playing instruments, listening to music, or making music with others may be highly reinforcing and motivating for some students.

Music therapists can utilize various methods to promote positive behaviors, including the following (Adamek & Darrow, 2005, pp. 188–189):

- Creating a sense of predictability and routine in the session, such as using a visually prominent schedule board, having a predictable room set-up, or using familiar materials.

Mrs. Hawkins' music therapy space is a model of predictability. When the children enter the room, they know exactly where they will find the rhythm instruments, the chairs, and the music board, which spells out the order of musical events for each day's session. Today, the board presents a sequence of pictures that represent (1) the opening song/warm-up, (2) a movement activity, (3) time for singing, (4) playing instruments, (5) surprise to allow for teaching flexibility and student choice, and (6) closing.

Everyone thought James would do well in regular music classes. However, his need for consistency and difficulty with change become clear on the first day. Mrs. Lin, the teacher, looked at the clock and said, "Let's sing

> 'Five Little Monkeys Jumping on the Bed.' We have just enough time for 2 little monkeys to jump on the bed today." James's response was a major tantrum with screams of "No!! There must be 5." That's because every time he sang this song, they started with the number, 5. Only 2 monkeys! That was just more change than he could handle.
>
> Mrs. Nelson, the music therapist, was contacted to set up a series of sessions to work with James on coping with a change in routine. A "surprise" symbol, selected with the ASD classroom teacher and speech clinician, was introduced. Visuals with choices for changes in how the music was performed including fast/slow, loud/soft, or instrumentation were made and familiar songs were sung with changes. Next, dice were presented and James practiced rolling, saying the number and setting out a matching number of items. "It's a surprise!" was said frequently when the dice were rolled. James was ready to choose a song, roll the dice, and sing using the "surprise" number. A social story song about "No Turn Today—but it's OK" was sung to the numbers that were not chosen.

- Practicing flexibility or tolerance of change by varying a musical element within the song or activity.

  > During instrument playing, Mrs. Hawkins integrates small but tolerable changes in the routine by having the children play a favorite well-known rhythm pattern, but then adding a rhythm instrument different from that originally taught.

- Providing positive reinforcement for appropriate behaviors.

  > As Jamilia keeps her hands to herself in the "Little Red Caboose" song, Mrs. Hawkins praises her for good bubble behavior, and she asks her to choose which instrument she would like to play during instrument time.

- Providing a means of communication for the child, and making sure that communication system transfers to all environments, including music therapy.

  > One of the children in Mrs. Hawkins' ASD group uses basic sign language, while another uses visuals to communicate. Mrs. Hawkins makes sure that she utilizes of these communication systems in song time and all other aspects of her session.

- Teaching peers how to interact in a positive way with the child.

  > Before Jamilia started to attend the regular music class, Mrs. Hawkins sat down with the children to discuss practical ideas for getting along with

their classmates, such as moving a little bit further away if a classmate moves within your bubble space.

Music Therapy to Improve Academic, Physical/Motor, and Leisure Skills

In addition to the aforementioned areas of difficulty for children with autism (communication skills, social skills, and behavior), other areas of deficit such as academic skills, physical/motor skills, and leisure skills can be addressed through music therapy. Studies indicate that music can be used to enhance memory, attention, executive function, and emotional reasoning (Bettison, 1996; Hermelin et al., 1989; Ma et al., 2001; Reitman, 2006; Thaut, 1988; Wimpory et al., 1995). Figure 2 shows examples of how music therapy can promote preacademic and academic development.

Music as a carrier of nonmusical information
- Songs with lyrics about world facts, math, vocabulary, body parts, and other academic content.

Music listening to promote a learning environment conducive to attention and focused learning (Holland & Juhrs, 1974; Litchman, 1976).

Music as reinforcement (reward) for compliance in academic tasks

Music to teach specific concepts
- Music interventions that require use of numbers
- Music interventions that require following multi-step directions
- Music interventions that present colors, shapes, forms, and other concepts
- Music interventions that require practice of auditory memory or auditory motor memory

Figure 2. Examples of Music Therapy for Preacademic and Academic Development (Thaut, 1999)

Songs, chants, and rhythm activities can be used to reinforce math skills such as counting, 1:1 correspondence, and ordering. Songs with repetitive lyrics, rhythmic patterns, and added movements can give children many opportunities to count, add, subtract, and place items in order in a fun and interesting way. Categorization by size, shape, color, and sound can all be practiced through simple music activities and musical instruments. Other concepts such as high and low, in and out, front and back, slow and fast can be taught and practiced through music therapy interventions combining music and movement. Incorporating movement along with songs and rhythm gives the children different options for understanding and remembering the concepts, making the concepts come to life through music and movement. Music

CASE TWENTY-SEVEN

Integrated Music Therapy With A Schizophrenic Woman

GABRIELLA GIORDANELLA PERILLI
Psychotherapist - Music Therapist
Secretary: Italian Society of Music Therapy
Rome, Italy

Abstract: *A comprehensive approach to music therapy was taken with a 20-year old woman diagnosed as residual schizophrenic. The 18-month treatment period was divided into four developmental stages in order to formulate goals and develop the most effective strategies. Techniques include: singing and playing pre-composed pieces, action songs, improvisation, song-writing, and projective listening experiences.*

BACKGROUND INFORMATION

At the time of therapy, Mary was twenty years old, and had been diagnosed as having *residual schizophrenia* in remission, with no organic damage. She was taking a minimum dosage of Serenase, a *neuroleptic drug*.

Mary was her mother's fourth child, and her father's first. Her mother had already had three sons from a previous marriage; her mother's first husband had died. Mary was born fifteen years after her three brothers.

As a child, Mary had disturbing dreams which caused her great anxiety. She also had early eating disorders. At school, she demonstrated several problems, including deficits in attention and memory, and hyperactive behavior. When she was three years old, she broke one leg while playing with another little girl. This turned out to be quite a significant event, not only because it frightened her so much, but also because she did not grow or gain weight for one full year.

Her family were both overprotective and critical of her at the same time. They were also of the opinion that it was impossible to cure Mary's mental illness, or to modify her disruptive behavior.

At the time Mary was referred to music therapy, she lacked many adapative, self-help and social skills, and had poor contact with reality. She had been in verbal psychotherapy for eight months, receiving three session per week. Nevertheless, her condition seemed to worsen, and her psychiatrist and parents decided to refer her to music therapy to see if a nonverbal approach might have an effect.

Mary demonstrated various forms of thought disorder, including *delusions*, fantasies, *loosening of associations*, poor reality-testing, and difficulties in concentrating, thinking, and remembering. She was unable to stay on-task for sufficient periods, and shifted topics of conversation from one moment to the next. Generally speaking, any kind of ordered behavior posed a challenge. Mary also had difficulty making decisions, not only because of her disturbed thinking, but also because she lacked interest and motivation. She was filled with ambivalence, and she rarely derived pleasure from anything she did. Her affect and mood ranged from *dysphoria*, depression and anxiety, to agitation and hyperactivity. She cried frequently. Mary also had disturbing nightmares, and a decreased need for sleep. Her sense of self was also disturbed. She felt worthless, and she often yelled out self-criticisms. She was generally withdrawn from others, and avoided interpersonal contact. She often exhibited hostility and anger towards others.

METHOD

Given the wide variety of problems that Mary presented, and the etiological complexities of schizophrenia, it became essential for me to clarify my own theoretical position and methodological approach as I began to work with her. We know that the schizophrenic process distorts the person's entire reality system, and disrupts important links between the ego and outside world (Andreasen, 1985). We also know that one of the most devastating symptoms is the deterioration or inadequate development of basic psychological functions, including those pertaining to thinking, affect and interpersonal relationships (APA, 1987).

I have found that many of these psychological functions can be approached in an integrated fashion through cognitive methods of psychotherapy. The works of Ellis (1962) and Kelly (1955) are particularly relevant. Taking Ellis' perspective, Mary presented two irrational beliefs or dysfunctional ideas: (1) "I am an incompetent person," which leads to feelings of inadequacy, depression, and lack of pleasure; and (2) "I cannot bear it," which leads to anger, hostility, aggression and inactivity. Taking Kelly's perspective, Mary does not have an adequate *"personal construct system"* for adapting to the world and changes therein; she therefore has an ongoing dread of what the future may bring.

From a cognitive point of view, Mary's problems formed a vicious circle: (1) She was terrified by demands from the environment, such as "You must be good in order to satisfy others," or "You must go to school and do well." (2) This led her to react with anxiety and self-deprecation: "I have no control over what happens to me;" "I can't do what everyone expects;" "I am incompetent, no good;" "I hate myself." (3) With these ideas, she would fall into depression: "What's the use, why bother to do anything?" "Life is miserable and meaningless;" "I can't enjoy anything." (4) The depression then led to blame and hostility towards others: "It's not my fault, it's yours;" "You are bad and incompetent." "I am angry at you, leave me alone." (5) Then, suffering the consequence of her inadequate behavior and the reproaches of her parents, Mary would feel punished and powerless: "I cannot do anything to stop what is happening to me." (6) This completed the circle in that she now would become terrifed by demands being placed upon her.

When I work with psychotic patients, I divide the therapy process into four stages. With Mary, these stages helped me to formulate goals and to devise appropriate therapeutic interventions. The first and second stages are considered *rehabilitative* in nature, and third and fourth are *reconstructive*. The main priority for the first stage is to establish contact and to gain rapport with the patient. The second stage is aimed at stabilizing current adaptive functioning while also restoring previous levels. In the third stage, specific problems are targeted for resolution. In the final stage, efforts are made to restructure the personality so that therapeutic gains can be maintained, and some degree of independence can be established.

Because my aims in therapy address the whole person, and integrate cognitive, affective, physical and interpersonal functions, I use a variety of music therapy techniques, as also recommended in Unkefer (1990). In my work, I integrate active and receptive techniques and use both structured and free musical experiences. Some are creative and some are re-creative. Some require perception, others require projection. I also shift the amount of interaction that is required. For these reasons, I call my approach integrated.

Using the full gamut of music therapy in this way allows me to gear the techniques to each stage of therapy and its corresponding goals. In this way, the client's needs are kept in the forefront. The following are the main techniques I used (and modified at the different stages) with Mary: playing and singing pre-composed music, action songs, improvisation, song-writing, and projective listening activities.

TREATMENT PROCESS

Stage One: Contact

For the first three months, I met Mary four or five times per week, each session lasting one hour or more. I felt that we needed to have frequent contact because of her disintegrated state. The sessions were held in my private office, a room large enough to accommodate Mary, my assistants (a musician and a social worker) and myself, and one that would permit free body movement. Most of the sessions were tape-recorded.

Establishing contact and gaining rapport with Mary was quite difficult. She did not like her previous (verbal) therapist, and in fact, described him as "awful." She experienced him as being very frightenining and intrusive, and even had nightmares and fantasies in which she imagined that the therapist was threatening to terrorize or kill her.

So at the beginning, I tried to approach Mary on neutral ground and on equal terms. I wanted to lower her anxiety level by respecting her deep inner space, and by not asking her questions about her maladaptive behavior, her past life, or her present life problems. I avoided any talk about her small stature and how it made her feel, her particular behavior towards people, and her hostility at home. In musical experiences, I was very accepting and nonjudgmental of everything that she did, and I found that this approach greatly increased her willingness to work on her own problems.

In our improvisations, my role was: to mirror her musical behavior, to give her feedback so that she could hear her own music, to help her stabilize and control the various musical elements, to encourage her to reproduce patterns, and to help her to express her negative feelings through music. I also improvised or composed songs with lyrics aimed at giving her an opportunity to become more aware and accepting of her body (e.g., I'm playing the drum, while touching my head, I'm playing the triangle while touching my arms, I'm playing the harp while moving my body, etc.).

We often had rhythmic dialogues. Mary often entered the room confused, agitated, and anxious. On one occasion, after finishing the welcome song, I asked her to play a duet on the drum. I began to play very ordered rhythms in 4/4 meter, and Mary responded with incoherent and disorganized beats. I let this continue for about three minutes, and then suggested that we take turns leading and following. She agreed, and as we went back and forth, her playing became stronger, more ordered and responsive. Patterns began to appear. Whenever we would match each other accurately, Mary smiled, and this seemed to relieved her tension a great deal. We ended with calming rhythms, and then listened to the tape-recording of our duet. I asked Mary what she thought about it, and she said that she was satisfied. We then talked about the process of making music, and I pointed out that first comes confusion, then attention to what is happening, then organization of ideas, and finally interaction. By this time, Mary was much more relaxed and was able to not only understand what I said but also repeat it.

Mary also did solo improvisations on various themes. In one particular session during this stage, Mary improvised two very different pieces on the piano. The first one was organized and coherent, and after hearing it, she said that its theme was "The Death of My Piano Teacher." During the improvisation, she demonstrated very good concentration, and was in contact with her ongoing feelings of sadness and discomfort. The improvisation was a study of the relationship between the various music elements: the use of the rest as a moment of reflection and an opportunity for progressing and developing her ideas in a logical and consistent manner; collaboration and integration between the right hand and the left hand; the use of regular rhythmic accents; and a perfect synthesis of thoughts and feelings within musical creativity.

The second improvisation was characterized by a lack of organization and coherence. This theme she decided in advance: "The Happy Country Woman." In this one, there was an absence of awareness and reflection; the music was confused and she made little effort to relate ongoing ideas to previous ones. Her concentration was poor and she did not pay attention to details such as accents, meter, rhythmic patterning; there was no unifying idea or sound quality; there was no homogeneity in the expressive material, and as a result there was no emotional coherence; momentarily, Mary would express her obsessive thoughts and anxious feelings through perseverative playing; there were no moments of rest or relief; and when dissonances occurred Mary seemed not to know how to resolve them. Overall, the improvisation sounded like a spasmodic search for something unknown which eventually began to bore her.

Based on the feelings that I experienced as I listened to Mary's improvisation, combined with a cognitive analysis of the musical structures themselves and Mary's verbal self-deprecations (I am stupid, I do not have any ability, etc.), I came to a better understanding of her inner state. Coexisting were intense feelings of depression and a pervasive confusion due to ego anxiety. These feelings of depression and confusion formed the basis for planning sessions for the next stage. I wanted to organize her personal construct system, lower her anxiety, and let her experience positive emotional states.

Stage Two: Stabilization and Rehabilitation

Goals for this stage were: to help Mary to focus her attention, and especially in tasks requiring perception and memory; to lengthen her in-task behavior and attention; to increase goal-directed behavior and perseverance; and to help Mary gain greater awareness and acceptance of her body.

To accomplish these goals, I used a variety of structured listening and imitative tasks. These included: having Mary compare two musical patterns and describe how they were similar or dissimilar; having her imitate rhythmic and melodic patterns that I presented; having her vary the same patterns in systematic ways (e.g., louder/softer, faster/slower, louder and faster, etc.). Two kinds of improvisations were used: ones that focused on here-and-now experiences, and those that explored a particular feeling or issue. Mary also told stories and drew to different pieces of music and made connections between them. Finally, we played a musical game which combined singing and matching instruments to various body parts.

The following is a transcript of a session which typifies this stage of therapy.

> *Mary arrives complaining that she is unable to think. She feels terrible because this prevents her from being able to understand when people talk to her. She also gets confused about how to do things. Her face shows how tense and upset she is.*
>
> *Given her state, I try to comfort and calm her through a listening activity. I ask Mary to tell a story, and she begins: Laura is a young woman who lives in a castle; she is in the garden taking care of the flowers. There are animals and birds in the garden. She is somewhat uncomfortable there because it is not as orderly and calm as she would like it to be.*
>
> *Two Princes arrive, and Laura joins them in the castle. She tries to explain to them how desperate she feels because she cannot take care of the flowers by herself. They quarrel. The Princes are eventually persuaded to cooperate with her, and they prepare lunch and all eat together. I help Mary to finish the story, and then wonder to myself whether I should proceed in a structured or unstructured way.*

I decide to invite her to select instruments to describe the main elements of the story. After making her choices, she begins to improvise on the drum (representing the castle) and plays a strong, steady beat. She then goes to the cymbal and sistrum (representing the Princes) which she plays in a loud and confused way. Changing to the autoharp (representing Laura), Mary begins to play soft, delicate arpeggios, followed by a steady beat on the hand-drum (representing animals), a simple rhythm on the triangle (representing birds), and scales on the xylophone (representing flowers). She then returns to the cymbal and sistrum and resumes playing in a loud and confused way (Princes). I then took up the autoharp (Laura), and accompanied her with the delicate arpeggios until she ended.

I then asked: What is happening to Laura now? Mary replied: "She feels happy, but still worried that she will not be able to find herself." Surprised by how personal this statement is, I asked Mary what she could do to help Laura. Mary answers: "Sing a song." She begins the lyrics with "Laura is unable to take care of the flowers." I then question her, "Will these words help Laura?" Mary replies negatively, and I ask "What would be more helpful to tell Laura?" Mary begins again and sings: "Laura is able to take care of the flowers." Feeling more comfortable with this solution, I begin to sing the same words with her, and as we repeat the verse over and over, I ask her to dance together.

As we finish, I ask Mary what she thinks about her song, and she replies, "Nice!" I then ask her how Laura feels, and Mary says: "Happy... because she now understands that it is her task to take care of the garden." I then ask if Laura will be able to find herself doing that, and Mary replies, "Yes, I think so."

I then invite Mary to play Laura's instrument (autoharp) and accompany herself singing the song. She decides that it is not the right instrument, and selects the xylophone instead. We sing the song again, with Mary playing a xylophone accompaniment. When I ask about it, Mary says that she is pleased with the instrument change. I ask her why and she says that she thought the autoharp was sad. I answer: "Sometimes I feel like that: the autoharp seems sad and quiet at times; it seemed more suitable to Laura when she was

> *talking to the quarrelling Princes, but when Laura is working, she is lively and strong." Mary agrees, and I comment: "Mary, you have thought about and understood what has happened to Laura in a very clear way. As you can see, when you decide to focus your attention, you can do it...you can think very clearly." Mary says, "Yes, I am really OK."*
>
> *I then begin to summarize: "You understand that Laura, the most important person in your story, was sad because she did not realize what she was capable of doing, but when she did, she became...." Mary inserted: "Happy, because she took care of the garden." I continued: Yes, she felt more self-confident because your song told her that she was able to do what she wanted." Mary agrees, and I ask her "Is it possible for you to say that you are able to do something?" Mary: "Yes, I am able to do something!" I ask her for an example, and Mary says that she can play something. I then ask her to say the full sentence, and she does: "I, Mary, am able to play the piano." As the session ends, I ask how she feels, and Mary smiles with the reply: "Better than before!"*

Stage Three: Problem Solving

By the third stage, Mary was more present in the here-and-now, and her cognitive functioning was much better. The goals were: to decrease her distractibility, to build problem-solving skills and logical thinking processes, to help her become more aware of her feelings in different situations, and to help her formulate some goals that she would like to accomplish with regard to herself (e.g., self-management).

Each session began with a welcome song, and after a discussion of whatever Mary presented that day, we went into either song-writing or tellings stories with background music. After this activity always followed a review and discussion of what she had done, and how she felt about it. I would then give her homework assignments which made her practice and put into effect things we had worked on during this session. (Prior to this stage, she was not ready for these homework assignments). The sessions always ended with a summary of everything she had accomplished and explored, and a good-bye song.

As evidence of her increased cognitive abilities, she worked on the same story over several sessions. It was called "The Vegetable Garden." When she finally completed the various chapters, Mary exclaimed: "This is my life!"

Stage Four: Reconstruction

In the fourth stage, the main concerns were: changing some of her irrational beliefs, improving her coping and social skills, clarifying her interests and needs, and trying to integrate various aspects of her personality.

Sessions during this period began with a welcome song, and a verbal discussion of any issues or problems on Mary's mind. Then, based on this discussion, I would engage her in one of the following musical experiences:

1) Composing lyrics and music to a song. This was done to work on her irrational fears or beliefs.

2) Playing or singing pre-composed music in different ways, or with different interpretations. This helped Mary to gain some insight into her own traits and characteristics when making music and in other daily situations. It also helped her to see that there were different ways to perceive and interpret things.

3) Improvising with or without a verbal theme, and with or without the therapist. These experiences were aimed at making Mary aware of how the music changed with the various feelings she was trying to express.

4) Projective listening activities (e.g., storytelling). These were used to help her integrate cognitive and affective components, and personal needs and goals through her musical experiences.

5) Listening to and discussing several musical pieces and then putting them into a hierarchy according to personal preferences or some other characteristic. This activity was aimed at helping Mary to adapt to changes in the real world, and to recognize how her own perceptions and preferences influenced her orientation to the world around her.

After the musical experience came a period of feedback and review. If Mary produced a song, improvisation or performance, I would tape it, and then we would play it back and she would react to it. Following this, we would talk about how Mary's music or the way she went about making it related to herself and her daily life. Based on this, I would give Mary a homework assignment. Usually, this involved rehearsing the songs she had written, and then using them throughout the week to guide herself into more functional and adaptive behaviors. At the end of each session, we would summarize what we had accomplished and close with a good-bye song.

The songs that Mary composed during this stage provide good examples of the issues and problems she confronted during this period. The songs were originally written in Italian; English translations are given below. Some of these were set to existing tunes (e.g., "I Accept Myself" was set to the Scout's "Farewell" Song); others had melodies especially composed by Mary.

I DON'T WANT TO BECOME REASONABLE

I don't want to become reasonable,
  it's too exhausting;
It's more comfortable not to work at it,
  anyway, I do not have the ability;
It takes time, time, time.

I decide to make an effort,
I think it's more convenient to have a better life;
I go step by step to be sure:
Exhausting obligation becomes joyful,
  Joyful, joyful.

SPRINGTIME TEARS

These are life's tears that I feel in myself
  I feel in myself.
When I am happier than now, the smile will come
  and glow again.
Joy gets down into my heart,
  Life creates love.
The joy will come from pain;
  soon merriment will come again.
I'd like to do easy things,
  within the need for toil.
I can try hard so that I am happier
Joy gets down into my heart
  And life creates love.
Joy will come from pain;
  Soon merriment will come again.

I ACCEPT MYSELF

I don't accept myself
as I'm afraid that other people don't love me anymore.
I am unsatisfied
because I think that I am unable to do anything.
But why? But why do I want to suffer?

I can accept myself as I am;
I am a worthwhile person, too.

DISCUSSION AND CONCLUSIONS

If we analyze the whole therapeutic process, we can see how music therapy played a positive role from beginning to end: it permitted Mary to overcome her resistance to therapy and to the therapists; it motivated her to play an active role in her own process of change; it helped me to reach Mary on the nonverbal level, avoiding the negative effects of verbal therapy; at the same time, it helped to integrate the verbal and nonverbal aspects of her problem, so that Mary could transfer the musical insights (i.e. irrational beliefs) and skills (e.g., problem-solving) outside the therapy session into her daily life (e.g., self-care, home activities); it allowed Mary to have multisensory experiences which helped to integrate fragmented aspects of her personality; it worked well when combined with other expressive arts to give Mary physical boundaries, and to move her from one kind of symbolic representation to another; it enhanced long term memory functions which were very poor; it increased Mary's symbolic cognitive abilities; it was useful in focusing Mary's attention when she was confused; it helped to modify some of her stereotypic behaviors; it provided a noninvasive way to address her needs, preferences, and goals; it gave Mary joy and pleasure and motivated her to be more playful; it facilitated social interaction, and lessened her feelings of inferiority; it provided tangible products that rewarded her hard work; and finally it provided an opportunity for her to develop a sense of pride in herself, and especially when sharing her musical accomplishments with her parents.

The advantages of music therapy are further underlined by how it compared to previous treatment using a verbal psychoanalytic approach. In response to the verbal approach, Mary lost contact with reality, and her condition worsened. This did not occur within the music therapy, except when sessions were interrupted for some reason (i.e. summer holidays), or when I inappropriately evaluated Mary's readiness and introduced verbal interventions prematurely. Clearly, Mary had difficulty with a verbal approach, and whenever she was not ready to approach her problems at the verbal level, it was necessary to return to music therapy activities alone.

Throughout the various stages, Mary's music changed considerably, going from confused to ordered, and from being disengaged or disinterested in the musical process to recognizing how her own feelings and personality were reflected in it. Her participation was at first whimsical, later music was a means by which she could enter into a personal problem-solving process, either alone or with the therapist's help.

My evaluations of her progress in the area of adaptive behavior were later confirmed by objective observations made by her parents and by Mary herself. Also, her psychiatrist reduced her medication to a very minimal dose.

Mary's social and coping skills are now quite adequate, and permit her to be involved in more satisfactory relationships with others. She is also more self-reliant, and has even decided to attend a pottery school. She is able to do a little cooking, and she now enjoys taking part in parties, trips and various recreational activities. Her thinking follows more logical sequences, and this permits her to follow a conversation, story or movie, and to express her opinions and emotions more easily.

Although I have presented the treatment process as if it occured in four clearly defined stages; in actuality, sometimes a session would contain elements of one stage exclusively, and at other times, a session would contained elements of several stages. This would depend on how Mary's readiness changed from one week to the next. Thus, during each session it was possible to consider peripheral or central issues, more specific or general ones, and to use simpler to more complex procedures. It was also possible to use a technique for assessment purposes at one time, and then later for treatment and evaluation.

In some cases, the same technique (e.g., music combined with other expressive arts) was used at two different stages (e.g., first to develop particular cognitive functions [perception, memory], and then later to analyze thinking modalities (e.g., loosening of associations) or to learn a problem-solving skill. Moreover, techniques such as song-writing were useful not only during the therapy session itself (e.g., to express and overcome a depressive state), but also as a homework assignment (e.g., to rehearse more functional self-statements or to develop a new self-management skill).

It is important when working with individuals like Mary that the therapist be always present as a guide and/or partner. The therapist must always have a flexible attitude, and be ready to modify the planned program or session to meet any needs that arise unexpectedly or to accommodate any shifts in the person's readiness to forge ahead in the therapeutic process. I have also found that having assistants was very helpful, and in some cases, essential. They were particularly important when we would play out a story with many characters, as their personalities would help to teach Mary about the complexities of social interaction, and the many aspects of her own personality that she could call upon in various situations.

In conclusion, I believe that integrated music therapy achieved such good results with Mary because it is a nonverbal, multisensory, joyful approach. As a result of the various techniques, Mary is now more aware of herself in relation to the real world; she can integrate different aspects of herself; and she is better able to verbalize her needs, resolve conflicts, process information. Now she is discovering the meaning and purpose of her existence, and has begun hear a new melody of life: "I accept myself as I am ---I am a worthwhile person, too."

GLOSSARY

Delusion: A fixed, false belief. A delusion involves believing things that are not real, and are frequently persecutory, grandiose or somatic in nature.

Dysphoria: A mood state which causes discomfort (e.g., anxiety, depression).

Loosening of Associations: Disorganized thinking and speech.

Personal Construct System: A psychological concept developed by George Kelly (1955) to describe how a person construes the world, and organizes his/her constructs into themes which permit anticipation of future events, and prediction and control of interactions with other people.

Reconstructive Stage: A period or phase of clinical treatment aimed at producing cognitive, emotive, or behavioral changes needed to achieve a more integrated personality.

Rehabilitative Stage: A period or phase of clinical treatment principally planned to regain previous levels of cognitive functioning or reality adjustment.

Residual Schizophrenia: A type of schizophrenia in which there are: a history of schizophrenic episodes with psychosis prominent, a current clinical picture without any psychotic symptoms, and continuing evidence of illness, such as blunted or inappropriate affect, social withdrawal, eccentric behavior, illogical thinking, or loosening of associations (APA, 1987).

REFERENCES

American Psychiatric Association (APA) (1987). Diagnostic and Statistical Manual IIIR (DSM IIIR). Washington, DC: Authors.

Andreasen, N.C. (1985). The Broken Brain: The Biological Revolution in Psychiatry. New York: Harper & Row.

Ellis, A. (1962). Reason and Emotion in Psychotherapy. Secaucus, NJ: Lyle Stuart.

Kelly, G.A. (1955). The Psychology of Personal Constructs (Volumes I and II). New York: W.W. Norton.

Unkefer, R.F. (Ed.).(1990). Music Therapy in the Treatment of Adults with Mental Disorders: Theoretical Bases and Clinical Interventions. New York: Schirmer Books.

Voices: A World Forum for Music Therapy, Vol 11, No 1 (2011)

Performance in Music Therapy: Experiences in Five Dimensions

By Peter F. Jampel

Introduction

Building community through music therapy performance has been at the heart of the work that I have done with a band of musicians who have serious mental health issues over the past twenty years at the Baltic Street Clinic in Brooklyn, New York. The complexity of their personalities as expressed through performing music intrigued me and challenged me to try to understand how to best work with them. This examination led me to ponder not only what worked and what did not in terms of treatment strategies, but also how it worked and why, questions that are critical in addressing what is therapeutic about performing music when working with people who experience persistent mental illnesses. Eventually an approach of identifying and treating these issues developed. In this process, I have considered the thorny question of how to promote the health of the individual performer, work with the manifestations of their illnesses while also attempting to build community.

This paper will address performance from a music psychotherapy perspective something that in the current context of community music therapy literature, is controversial. Additionally, it discusses the process of performing music in terms of dimensions of experience. This approach allows the clinician to describe, assess, analyze and evaluate the components of what is happening both over time and in the moment. It is my intention to develop a language not only *between* therapists but *with* clients. It is an attempt to promote a consistent means of description. The five dimensions of performance in music therapy are not meant to exclude other possible dimensions that might exist but were found to be a concise, descriptive shorthand that covers the essential aspects of the performance experience (Jampel, 2007).

Theoretical Perspectives on Performance in Community Music Therapy

Communities unlike the one at Baltic Street that are formed through shared neighborhood, religion, hobbies or employment have the advantage of the cohesive effects of having things in common, of membership that reflects a willing intention to be part of community (Stige, Ansdell, Elefant & Pavlicevic, 2010). Bonds built on the basis of having joined a community because a mental or physical illness has seriously impaired one's lifestyle or level of functioning, do not necessarily reflect shared beliefs and interests. These membership conditions of necessity can contribute to a sense of alienation, of having been thrown together with other people because of the unfortunate circumstance of an illness that itself, can be hard to accept. In order to promote community connection and belonging that is heartfelt, people in this instance need to find some greater sense of meaning and purpose. How can music performance foster a sense of volitional identification, strengthen the bond of connection, and promote a feeling of purposeful action?

The community discussed in this article initially reflects the kind of thrown together quality that Gary Ansdell (Stige et al, 2010, p. 44) discussed as *circumstantial*. As the music therapy program expanded from its origins in individual and group music psychotherapy, an environment was created which provided increasingly frequent opportunities for clients to perform music at a monthly cabaret called *The After Hours Club*. This seemed to strengthen the bonds not only between individuals who played music but also collectively among those who came to listen. Ansdell explains this process as communities forged through *communication and practice*. They increasingly share music as a way of communicating. It serves "to actively construct, sustain and develop particular modes of community, and its accompanying experience of belonging" (p. 47). Music performance groups in my experience are more motivated to learn how to communicate in music because such skills improve the quality of performances. In this view, communities are created through the developing competencies and shared interests of its members. Practice and rehearsals consist of doing and learning something together repetitively, through the experience of sharing passions, interests and knowledge, of planning together while learning to negotiate differing musical needs and tastes. Through this process, individuals can experience meaning, identity, engagement and ultimately belonging. An essential characteristic of this type of community is the acceptance of the difference, strangeness or otherness of each member of the group. Stige (2002) call this "unity beyond uniformity" (p. 173). Music making provides equipoise between the individual's state of existence and those groups to which they belong.

The achievement of a balance between individual needs and the good of the group is no small feat. In my experience, considerable skills are needed to prepare the individual musician to be able to contain personal agendas and to come to see the larger rewards of sensitive listening and playing. Central to this process is preparation to perform. For some music therapists like Stuart Wood (2006), preparation for performance involves individual sessions and music therapy workshops in order to create a psychological foundation for building meaning and readiness. His work with people who have experienced neurological trauma addresses the differing demands made on clients in stepping from private music making to the public nature of performance. His matrix model of community music therapy customizes the design of music therapy services around the particular needs of the individual. Though we work with different populations, I have also found that services that are designed around the special needs of the individual tend to promote a better balance between those personal needs and the context of working within a community.

It is of utmost importance in my experience to design music therapy services that take into account individual complexity along with the psychological ramifications of performance in order to avoid possible undesirable outcomes. Inadequate preparation can result in feelings of stress, anxiety or even worse, a sense of failure. This perspective on performance preparation or the lack thereof is offered by Jon Hawkes, a well-known Australian cultural analyst, in an interview conducted by the music therapist Katrina McFerran (O'Grady, 2008). Music performed publically, he warns, risks one's health by the effects of the adrenaline surge that often accompanies it. This he believes accounts for the prevalence of performance anxiety. He encourages music to be made rather than witnessed and goes on to discuss the culture-bound aspects of performance as *music making* versus *music witnessing*. In Hawkes' view, performers who perform with and for one another learn to direct their energy (and satisfaction) in connecting to each other and not to their audiences. He believes that only when performance emerges as a genuine client interest should it be then pursued as a possible direction by the music therapist.

Another perspective on the balance between the experience of the group and that of the individual is offered by Ansdell (2005, 2010) as he addresses performance as both a *self and collaborative* effort. In his view, reparative work is done both within the individual and to their connection with others in both spheres simultaneously. Careful attention to musical listening and organization of material in rehearsals is needed as clients are assisted to work through pathological attitudes in their self-identity, social relationships and work lives. He maintains that the collaborative nature of performance and the public completion of this process act powerfully to contribute to the meaning derived. Ansdell does recognize the pressures that can exist on performers when the experience is ladened by attitudes of competition and judgment. This position resonates with my own observations about the mutuality of growth and development that can occur for individuals within the context of a maturing performance group.

Although Ansdell and Wood seem to appreciate the potential pitfalls of performance, what is absent in their discussion is some systematic way of assessing who their clients are as performers, what clinical issues they present for treatment and how performance can address them. Without such procedures, I have experienced the discussion around individual cases to lack clarity and specificity. In his work with his client Maria Logis, Alan Turry (2005) addresses these challenges. Finding meaning in performance, carefully working through the implications of this process for the client, and understanding its impact on the therapeutic relationship are all hallmarks of Turry's work. Through the process of song improvisation, she went on to become a songwriter, performer and inspirational speaker sharing with her audiences the power that music and performance had on her struggles to survive cancer. Carefully sifting through her transference and his own counter-transference, Turry explores the implications of bringing these improvised songs into the public domain. He develops a procedure to assess and ascertain information about her capacity to form trusting, intimate connections through performance while also listening for the possibility of substituting external recognition for authentic relationships.

These theoretical perspectives illuminate the essential challenges that I have experienced. The ensuing discussion will develop the historical context in which they came to life.

Background

The people I worked with are adults diagnosed with mental illnesses such as schizophrenia, bi-polar disorder, severe anxiety disorders and recurrent depression. They are people from a diverse, heterogeneous ethnic, cultural and racial background typical of clinical populations in New York City. This cultural context is critical when considering the attitudes I encountered about performance. This was reflected by the many styles of music that were made, the attitude toward solo versus ensemble performance, and the expectations that these musicians have with and from their audiences. These embedded cultural aspects of music performance are very particular and vary significantly from that of other cultures (Nzewi, 2006; Oosthuizen, Fouché, & Torrance, K., 2007; & Inoue, 2007). The individuals discussed here are those drawn deeply to making music. This is a subset of the general adult psychiatric music therapy population. Their love of music is far more critical than the extent to which they are gifted. Almost invariably they share a sense that making music is a core aspect of their selves.

It is not my intention to imply a hierarchy here where performing music is somehow viewed as the pinnacle of therapeutic music making. It is just one possible direction but one that seems more likely to occur when music therapy is conducted in a long-term community setting such as the one described in this paper. In this context, wanting to perform can be seen as a natural extension of the evolving interest that can take place in community music therapy settings (Aigen, 2004; Jampel, 2007).

Though there was interest in performing music from the inception of the music therapy program that began in 1975, there were also challenges to address. The performers who emerged in the community sing that ended each week's program had problems ranging from insecurity and musical inhibition to narcissistic exhibitionism. Singers, bass players, drummers and the occasional keyboard player or solo instrumentalist would emerge and gain recognition for their emergent talents. Even though these musicians solidified into a back up band, it was difficult to find a consistent time and place to work on the problems they experienced in performing music.

By 1991 with the addition of a monthly cabaret called the After Hours Club, they had coalesced into the in-house performance group that has become The Baltic Street Band, a group that continues to this day (Aigen, 2004). The members of the band wanted more time to play with each other. They were motivated to accept more discipline as their interests in music performance grew. This process became more gratifying as their labors produced music that gained increasing recognition from their peers. New members joined the band as their exposure grew. Jobs developed by playing for other communities and some of these were paid performances. Today the band continues to both perform and record its own music. Yet as it became bigger and more skilled, personality issues became increasing obstacles. Lack of preparation, lateness, missed practices and ongoing conflicts between certain band members became disruptive and divisive.

My perception was that persistent personality problems interfered with the working environment of band rehearsals. Patterns emerged in the clashes between performers that appeared to be repetitive and pathological. Patterns of disturbance became evident regarding certain performers' relationship to their own music making process. Sometimes these issues seemed to be complicated by the presence of an audience. Interventions designed to address these issues were made during rehearsals but often the resulting process took time away from preparation for performances. This led to resentment from some musicians that valuable rehearsal time was being consumed by extra musical difficulties. Not to address these recurrent problems seemed out of character with my intention of promoting healthier group dynamics and facilitating individual growth. I perceived that the flaw in this strategy was not in my therapeutic intention but in trying to do both a therapy group and a band rehearsal at the same time. Neither was being done as consistently nor as effectively as was needed.

The Music Therapy Performance Group (MTPG) was first designed in 2004 to both research and treat this dilemma. Meeting as it did on the morning of rehearsals, it was intended to siphon off the need to make therapeutic interventions during band rehearsals. The group focused on the needs of the individual performer. It did not take time away from performance preparation but was in addition to it. Attendance was encouraged but not required for band members. Later after the research study was completed, another treatment group was formed in 2006. This group was opened up to other musicians who did not as yet feel ready to perform. Rehearsals were now free to get down to the business of making music in the time frames needed in order to get ready for upcoming gigs. Persistent and distracting pathological behaviors such as chronic lateness, unexplained absences, lack of work preparedness, or seemingly intractable personality conflicts were addressed in the MTPG. It was here that we could consistently work on performing music and the personality of the performer.

Assessment, Treatment and Evaluation Considerations

Interviews were conducted with each participant prior to the group. The interviews were designed to ascertain past personal history in music including the level of musical training each participant had. As a staff member, I had access to medical records that contained other aspects of their past personal, medical and psychiatric histories but music history was not part of the medical record. Additionally, information elicited in staff meetings from other members of the interdisciplinary team augmented my working knowledge of each member of the group.

Past personal history in music provides an assessment perspective that crosses an array of contributing factors: it provides a different glimpse into parent-child relationships and sibling rivalries; it allows the clinician to assess self-directedness, perseverance, and task completion; it offers a view on assessing the relative health of the creative personality from a viewpoint where spontaneous expressiveness is on one end of the spectrum and pathological anxiety, rigidity and narcissism are on the other; it also provides a window into cultural/family attitudes towards music and the effect this has on an individual's attitude toward making music. The assessment process takes into account the dynamics of musical activities in families where this is an important aspect of family life. Often times the expectations, musical accomplishments or disappointments of the parents are visited upon the child. These attitudes can be extended, altered or exacerbated by subsequent musical history with teachers, producers, promoters, musical juries or with other musicians.

The therapist should be looking to assess various areas of functional capacity. Does the individual engage differently in group music making environments than they do in one to one situations? Does the individual possess flexibility in their musical interactions with others or are they limited or inhibited by this? Does music highlight alternative areas for expression and learning that taps previously unknown or under-utilized pockets of intactness? This is particularly important for individuals who due to the onset of mental illness often suffer cognitive losses. Identifying past trauma in music making with parents, teachers, or in front of critics sensitizes the music therapist to the possible occurrence of traumatic re-enactment.

With sufficient knowledge of the person's musical background, the earned trust of the therapist, and the evolving safety of the group, conditions can now be established where promoting a re-working of the performance experience can allow for a greater sense of meaning and satisfaction to occur. The resulting treatment implications of this process point to the making of music in front of others as a necessary reconstructive strategy. Verbal interventions help create the corrective performance conditions and are also used to evaluate how effectively progress is being made. The relationship between assessment, treatment and evaluation are all bound together in a procedure that examines and analyzes performance from the five experiential components that will comprise the main focus of this discussion.

The Music Performance Personality Profile

This profile incorporates elements of the person's self-image as a musician and how this is embedded into the individual's overall personality development. This required assessing parental/family attitudes about music. In all but one case in the group, music was seen as a significant dimension in family dynamics. Often parents, siblings, grandparents, uncles and aunts were either musicians, musical or cared deeply about music. In some cases, the attitudes cited represented consistent and supportive parental involvement. For others, it was only a small island of support in an otherwise strained or neglectful relationship. Then for others still, past history of music demonstrated patterns of harsh and abusive behavior. With parents who were either musical or musicians themselves, aspects of the parent's past musical experiences often promoted a vicarious experience as the parent relived their own musical experiences through their children's accomplishments or lack thereof. Three group members had parents who were both musicians and abusive parents. For these participants, performance was an opportunity to try to understand and work through their still unresolved feelings about these troubled parents (Borczon, Jampel & Langdon 2010). Sometimes the histories indicated pathologically competitive environments in which siblings were compared to or pitted against one another in terms of the degree of perceived talent that led to preferential behavior. This often set into motion patterns of interaction within the group that seemed to recapitulate past family dynamics (Yalom, 1970).

When the assessment was completed, an initial plan evolved that targeted goals to build upon existing healthy ego structures and to identify pathological features in need of reconstruction. Patterns of behavior as they emerged within the group, were then brought to the awareness of the individual. The intention was to promote more accurate observing ego and the capacity for self-reflection.

Setting and Participants

The Music Therapy Performance Group occurred in a setting in which a multiplicity of music therapy services were provided including other group and individual approaches. One of the participants was 78 years old at the time and had been an active recipient of services at this facility since 1972. Many of the group participants had long-standing connections to the music therapy program. The long-term nature of their connection to treatment at this facility is vital in understanding the impact that the MTPG had.

The work described herein occurred over a period of two and a half years from 2006 to 2008. The new clinical group provided services to a total of 15 adults ranging in age from 24 to 78 years old. Five participants completed their participation prior to the end of the group. This was due to various reasons: changes in schedule due to work or school, moving away or ending treatment at the clinic or in one case, the completion of short term goals. The ethnic and cultural backgrounds of the participants varied. There were three people of African background (two from Caribbean origins and one African-American), five Hispanics, and seven white participants of various ethnic backgrounds.

Group size in any given week varied from as few as three to as many as nine members. They met once a week for forty-five minutes. Most of the group members were also members of the Baltic Street Band consisting of about twenty mentally ill musicians. The group met in the auditorium space where weekly rehearsals and monthly performances took place. The meeting space was in front of a small spot lighted stage located nearby to an acoustic piano. We sat in a circle in front of the stage. My antique tenor banjo sat waiting in its brown alligator and green velvet-lined case. One client loved to play it and his instinctive musicality allowed this to happen almost immediately. Guitars both acoustic and amplified, a bass guitar, hand drums, floor drums and a full drum kit were all made available to participants in order to facilitate the use of instruments for accompanying singers or for instrumental work. The stage, the stage lights and the instruments were all used to create an environment that replicated being in a performance.

Group Structure and Process

"So what would you like to work on in today's Music Therapy Performance Group?" threw the initial direction of the session onto the participants. Their self-image as musicians came up, how they worked with and reacted to other performers, and how they felt about particular audiences. This sometimes led to past performance memories and how their current experiences often had their origins in the past.

Once a theme was centered upon, we might process a recent performance experience including how they felt about it, what it meant to them, and what their reactions were about. For example, one group member who felt anxious and distracted by audience inattentiveness in the After Hours Club the day before, discussed her fear that nobody was listening to her because they did not like her singing. She recounted how in the past, she would become petrified to sing before her parents who she felt were dismissive of her singing when she was a child. After much encouragement from other group members, she sang the Cole Porter (1932) song "Night and Day" accompanied on tenor banjo by the group leader. This time with a new arrangement, she was able to sing in a more intimate way and connect more deeply to the song. The group response was warm and appreciative. This promoted her sense that she could re-construct the performance experience thereby making her less vulnerable to worrying about how she sounded to others. She developed a clearer picture about how the distraction she feared in her audience was really her own projection. By redoing this particular song, connecting more fully to the message in the lyrics, trying a different tempo, accompaniment and arrangement, she was able to alter her relationship to the music, her accompanist, the audience and herself.

We often processed how they experienced the music, how connected they felt to the song, how connected they felt to the other musician(s), how they experienced being listened to. This often brought up aspects of their pasts when similar feelings occurred and what this might signify. Repeating a song invited the musician to try to connect more deeply, concentrate more fully or to realize how the images that sprang to mind while performing might offer them additional fuel to work with. Explorations included: memorable past performances both good and bad; what life was like in their past performing lives; life on the road, special audiences, amazing performers they had seen or with whom they had played. They talked about drugs, money, jealousy, competitiveness, anxieties, self-image, sexuality, hopes, dreams and nightmares. The culture among performing musicians was a favorite topic. They talked of the camaraderie that developed between performing musicians. Their common fears, inattentive audiences, anxieties and insecurities but also the closeness forged by pushing through these difficulties and learning how to cope with them. While they acknowledged their appreciation for each other's skills, they also were able to acknowledge their envy and jealousy of each other as well. They learned by watching and imitating each other. Fine points such as intonation, breathing, microphone technique, body language and movement, story telling through song, stage presence, dressing and costumes, shaping your appearance, the use of make-up, and the capacity to read audiences all were frequent content areas in the group.

Performances were taken apart by the group and analyzed. Rivalries, fears and insecurities were processed but so too were triumphs. Feelings about the leader emerged, sometimes spontaneously, sometimes elicited. Often group members had powerful needs for attention or approval. If these dynamics seemed to be repetitive, the leader offered them back for individual reflection and feedback and as opportunities for the identification of triggering mechanisms. As pathological patterns among members of the group diminished, trust and cohesion developed which then allowed for more risks to be taken both in the music and verbally.

The five dimensional evaluation model was explained and used in the group to foster a systematic representation of performance as a dynamic psychotherapy process. The dimensions formed a shorthand language that objectified each individual experience into something that could be shared and comprehended. By exploring the family music history and the performance background of each group member, people came to a better understanding of themselves and of each other. It provided a focused context for personal disclosure that promoted a sense of safety and trust.

Attendance, Individual Communications and Transference

People who missed a session were expected to call. Unexplained absences or excessive lateness were addressed in the group. If it became a pattern, I met with the individual in my office. One member left the group due to his inconsistent and erratic attendance. He did continue to participate in the band. Several members experienced medical and psychiatric problems that kept them from coming in but as long as they kept in touch, they were encouraged to return as soon as possible. Three people were hospitalized during the course of the group, two for psychiatric reasons and one for medical reasons. Only one did not return to the group after his psychiatric discharge.

Attendance in the group was offered to but not required for members of the band. Eight of the active members in the band attended. Six members of the performance group eventually joined the band. The only one who did not was a nightclub singer who came to group after experiencing severe performance anxiety following a vocal lesson from a teacher who she felt was severely critical of her. This criticism brought up pre-existing areas of sensitivity that made her feel vulnerable and self-conscious. These conflicts illuminated historical patterns of criticism in her family. She was encouraged to try to perform in the group when she felt ready. After she finally did so, she reported that the support she received had helped her to resume her nightclub career as well as prompting her to find a new vocal coach. With this, she left the group reporting that she had accomplished what she had set out to do.

A number of people sought me out to talk to me about a range of issues outside of the group. I would see them but not to discuss group issues. If however I felt that a member was being disruptive to the group either by repeated unexplained absences or chronic lateness, I would request a meeting. I did this in an effort to provide additional support or clarification of problems that were occurring for them. This happened most frequently with two particular people who both had fragile and abuse filled histories. One person eventually left the group and the other Trisha, will be the subject of the case study discussed later in this article.

The issues brought up in group were considered confidential and group participants were asked not to discuss them outside of group. When necessary, I discussed significant developments both with the director of the band and with other clinicians on the treatment team.

I also attended the last hour of band rehearsal each week. I was thus able to observe how each individual operated. This gave me first hand information that I found useful in my work in the next weeks's group. I was an occasional instrumentalist and back-up vocalist in the band as well as the Master of Ceremonies for the monthly cabaret. In both of these roles I was able to witness their performances.

I played other roles outside of the group: administrator, clinician, and internship supervisor. I was also the individual therapist to three members of the group. These multiple roles had evolved over the many years that I worked there and tended to complicate my relationships in the MTPG. They became topics for discussion in the group and added complexity to the transference process.

One of my individual clients saw me as too busy and at times less than available to him. He dropped out of group after he exhibited inconsistent attendance but continued in individual verbal psychotherapy. He focused on his relationship with his deceased, abusive musician father who when not constantly working on the road, got into drunken rages at home where he threatened and abused his wife and children. He reported feeling that in the group, I did not having enough time to see him and that he stopped coming because of this. Over time, he became more able to discuss his feelings about my perceived lack of availability to him and established a better capacity to process his feelings. As he worked through his anger, his attendance in band rehearsals and in performances improved. The relationship that he had with his father affected his ability to connect with other musicians. The dimensions of how this process works will now be explained.

The Five Dimensions of Performance

The experience of performing music involves a complex interplay of connections: between the musician and the music that is being played; between performing musicians in terms of how they feel together playing on stage; between the performer and the audience in terms of the connection that develops between them in both directions; and between the performer and the thoughts within his/her own mind while performing. The experiential totality of these four co-existing states or dimensions, represent the presence of a fifth dimension - the feeling state of the performer. If all four previous dimensions are in a relational state of maximum connection toward themselves and each other, a complementary process ensues. The performer and the music are one, players riff off of each other and move more deeply into sync together, the audience gets drawn in by the action on stage which is felt by the musicians who then play off of the audience's energy, and the performers inner thoughts and feelings act to provide emotional connection which heightens the act of music making by adding depth and meaning. When all of this is synergistic, the effect can be riveting, transforming the moment into a *peak experience* (Maslow, 1971) or as Ansdell (2005) describes it *performance as epiphany*. Such reports described by the performers in the 2004 research group (Jampel, 2007) used the word "spiritual" most often to try to capture this electrifying sensation. But when one or more of these dimensions becomes disturbed, the result is a lessened sense of satisfaction. This can take the look of a musician who does not feel connected to the music, or performers who feel out of sync with each other, or an audience that by being noisy and distracted either affects or were affected by the performers on stage, or the inner thoughts and feelings of the performer (inner audience) becoming a source of negativity and loss of focus. All of this affects the fifth dimension. When looked at from this perspective, the goal of the music therapist is to identify areas of conflict within the four previous dimensions that may be inhibiting the performer's optimal sense of satisfaction and meaning.

These phenomenal states may be all operating simultaneously or in different combinations with each other. A performer can be so absorbed in the music that they might not be fully aware that they are performing before an audience. On the other hand, internal thoughts of doubt, insecurity or negativity may overwhelm the performer and promote a sense of isolation or alienation between the player and fellow performers or from the audience. Because it is a shifting and interactive process, evaluating the performance experience in this multi-dimensional model adds descriptive and causal flexibility. These five states can be best understood as a fluid experience with each dimension containing shifting phenomenological aspects. Repetitive patterns of disconnection in any one or more of the dimensions can be seen as performance psychopathology. Bad performances happen and the dimensional disturbances may be merely transient or reactive to recent stressors. But persistent patterns whose roots extend deep into personality formation tend to be more durable and difficult to work through. On top of this, one must account for the abnormal psychopathology states of psychosis, mania, or severe anxiety and depression that exist within this population. This often complicated the treatment process.

The evaluation process revolves around perceiving changes in any one or more of the dimensions and then adding to or modifying long term or short-term goals accordingly. This process entailed highlighting dimensional strengths and identifying and working through dimensional pathology. Self-reports and first hand observational data were the means by which this clinician was able to understand the impact that this approach had on each of the five dimensions. In the ensuing discussion, clinical vignettes will illustrate each dimension to be followed by a case study that will integrate all five.

The First Dimension: Connecting Within to the Music

Since the end of the MTPG in 2008, I continue to observe performers having difficulty connecting to their music. Whether it is a student, a client or a professional, one can at times detect through body language, facial expression or a perceived lack of conviction in the music that something is missing. Yet only so much can be known through observation.

Choosing the right material is a critical and necessary step. Some performers are clear about what they want to do, what feels right to them. They know themselves musically and when the time is right they may rediscover some old chestnut or to try something new or different. These individuals seem adept at interpreting their material and finding some deeper meaning that promotes connection to the lyrics or to the music. For others however, connecting on the first dimension brings up difficulties in making appropriate musical choices that seem to reflect on the larger problems they have in making appropriate life choices. They are unsure of themselves and of what may be expressive of their present mood. They lack conviction or passion about decisions, or are over determined to try something, or constantly change their mind or just give up too easily. Addressing this situation is tantamount to addressing issues such as why they cannot seem to find themselves, or not know what feels right for them, or how they continue to have persistent feelings of uncertainty about the direction they find themselves going in. Deeply connecting to one's own music can be seen as a statement about how well a person has learned to regulate life, of understanding what works best, of learning how to choose among the myriad options that life and music can offer and learning how to make the best decisions. Inside of this process, there is considerable room for making adjustments, finding new meanings and interpretations, looking and listening with fresh ideas and perspectives. If one cannot consistently regulate how to choose music satisfactorily, broader problems with life regulation often persist.

"What made you choose that song?" is a question that begins exploring the experience of internal connection to song choice. "How did that feel?" allows for the identification of associated affect after the song is performed. The crucial intervention is "Did that song seem right for you?" This question promotes an understanding about whether both song and affect were mood congruent. In other words, if the song evokes an unwanted mood or association even if the song had felt right at other moments, the song will start off as ill-suited for that person in this moment. Sometimes it may just be a matter of changing tempo, key, instrumental accompaniment or arrangement. These adjustments may allow the musician to feel the song differently and permit them to find an alternative path to get into it. But some pieces of music are just not right and if this is the case, the music therapist must facilitate that awareness in the client.

If left unresolved, the first dimension (D1) can spill over into dimension two (D2) and to three (D3). If the music does not feel right, playing it with others gets off track. At times, musical interaction may experienced as a beacon promoting one to come back. But if the music within is not sufficiently connected, the second dimension of connecting to other performers is often negatively impacted. This is I believe discernable to the audience. When disconnected to their own music and consequently to other performers, connecting to an audience is ever so much more difficult. The little voice in one's head experienced in dimension four (D4), can intrude into the performer's thoughts and wreck havoc with concentration and effort. Wrong notes, forgotten parts, overplaying can all result. This will all register in the performer's fifth dimension (D5) report of satisfaction and meaning. It is therefore most critical to find music that promotes the experience of inner connection. This process might be different with professional musicians who are required to play certain kinds of music whether they want to or not but in a music psychotherapy group such as this, this first element is vital. Knowing that you don't feel it but having to play the music anyway is one thing. Not knowing how to get that feeling is quite another.

A discerning performer may change or modify their play list depending on how they feel that day, or how they read the mood of the audience, or how illness or injury can affect song choice. An experienced musician uses these factors in calculating when to make certain adjustments. However, when these choices grow out of fear or anxiety about the music chosen, the result can reflect internal disturbances that can project unresolved D1 states onto the music. The question then becomes one of how to promote more accurate self-perception through greater connection to song choice.

Sometimes evaluating these factors is easier to do in a post performance process. After the dust has settled, an individual can often better assess how well they adjusted to the performance environment. Such discussions were frequent in the Music Therapy Performance Group. As trust developed, group feedback was more seen as supportive, non-judgmental and not as intended to be critical. Group members shared their observations with each other much like a Monday morning quarterbacking session where the team might look at the videotape (which sometimes happened) and discuss the performance from a detailed technical analysis viewpoint. Risk taking was promoted as performers encouraged each other to try out new ideas. However when this experience was perceived as criticism, the therapist had to carefully monitor this process and point out if old transferences were being kicked up. When things got to this point, often it was an indication that some pathological D2 aspects were emerging that required further exploration and understanding.

Though he no longer enjoyed playing music as a drummer, he was an enthusiastic singer and was also the one who loved to strum the tenor banjo. When given permission to sing, he was able to concentrate completely on his voice and enjoy himself again. This led to his performance in the After Hours Club several months later singing a duet of the song "Strangers in the Night" (Singleton & Snyder, 1966). He reported in the group the next day that he had felt relaxed and comfortable. To this observer, he looked and sounded fully connected to the music.

Dimension Two: Performers Connecting with Each Other

The conversation between musicians on stage is at its best a healthy, balanced and reciprocal relationship. There is a sensitivity in listening, selflessness, being together yet maintaining individuality, lifting and holding each other, a lightness, passion, a sharing of beauty and discovery. But when the musical conversation is a struggle, maturity is needed in order to resolve the inevitable conflicts that are normal even necessary in the creative relationship between adult musicians. Managing this phenomenon successfully promotes authenticity and healthy collaboration. But when patterns of conflict repeatedly emerge between particular performers, there is a likelihood of deeper disturbances dwelling within or between them.

According to the musicologist Christopher Small (1998) the meaning to be found in *musicking* is in the creation of relationships formed through the act of playing together not in the sound they produce. These relationships promote and nurture communication and listening skills and represent an ideal of how people work with one another. "Performance does not exist to present musical works, but rather, musical works exist to give performers something to perform" (p. 8).

Such was not the case initially for two members of the who fought incessantly with each other and who often but less frequently, battled with other group members. Performances for them were often highly mercurial events filled with behind the scenes drama that at times, spilled over into tantrums and sudden withdrawals. During performances these individuals often tugged in different directions from the other musicians on stage in terms of tempi, dynamics, pitch accuracy, feel and interpretation. Conflict always seemed to swirl around them. They collaborated with other musicians but could not work together. They did not seem to listen well and isolated themselves. They were easy to anger and to suffer emotional injury. Performance day reflected a high drama of uncertainty about whether they would appear at all or if they did, what they would perform and with whom. Trisha (our case study subject) would inevitably show up but late, unprepared and disorganized. She would often appear visibly upset during performances, grimacing and shaking her head while hurtling ahead of the instrumentalists ignoring cues and cadences. Another group member who was her primary antagonist in the band, would just not show up on performance day leaving fellow performers to scramble without his keyboard, drum accompaniments and vocals. These precipitous actions left the band shaken and mistrustful of both of them. Each had been in the original research group and both were encouraged to enter the new Music Therapy Performance Group.

Not surprisingly, these two individuals, one a 53 year old, white male, who I worked with in individual psychotherapy, and the other Trisha, a 62 year old white female, each had histories of being in abusive relationships. Both were diagnosed with Post Traumatic Stress Disorder. Both were the victims of abusive parents, both had abusive sibling relationships, and both had been involved in adult partnerships where patterns of abuse emerged. They also perceived each other as abusive.

The work for them in D2 involved uncovering the reasons for their tendency to be distrustful of others. Through a gradual process of disclosure of their family histories in music in the group, a context was developed between them in which they became better able to understand how they triggered feelings of being abused in each other. In fact a bond started to develop between them as they came to realize that they had much in common. They began to experience an empathic connection. They felt encouraged to try and sing together in group and when they did, both expressed satisfaction about how it felt. Finding a way to join their voices instead of using them to fight with each other helped to form a bond. They later decided to perform as a duet in the After Hours Club.

What unfolded for Trisha was a tendency to feel a sense of rivalry with others. She reported that positive attention in her family for her artistic achievements was hard to come by but for her pretty younger sister, it seemed to simply rain down for just being attractive. Criticism on the other hand seemed to pour down on her. She experienced this as ridiculing her efforts to sing or play an instrument. Her mother's critical voice often sounded within her head that she was just worthless, untalented and unattractive. For him, praise was just not available from his alcoholic and violent father. No matter how hard he tried, he and his efforts were belittled. Along with several of his eight siblings, music was a central

A MTPG member who had not performed for years reported he no longer did so because he did not like to play the drums. They brought up memories for him of playing in bands past where he felt too responsible to keep the beat.

aspect of family life. He performed with three brothers in a Rock 'n Roll band. This disintegrated into constant disputes, violent confrontations and episodes of alcohol and drug abuse although he reported that he never drank, smoked or abused drugs. Feuds persisted for years in this family. Music was the only means by which members of the family could come together though this was only a faint connection amid a sea of emotional family turmoil.

They fell into a relationship with each other where she thought that he was getting all the attention and he felt that her outbursts and tantrums were emotionally abusive to him. The group leader sought to bring their historical patterns of social interaction to their attention and by so doing, promote self-reflection and observing ego in terms of the causes and triggers for their behavior. Though his erratic attendance eventually resulted in his leaving the MTPG, he did become more consistent in attending After Hours Club performances.

She integrated the feedback in the group, learned to recognize similarities in the present to past patterns of abuse and victimization, and eventually opened up to partnering musically with him. They still had their moments together but each developed a greater capacity for connecting musically and emotionally. Both seemed less prone to missing cues, singing too loudly, making errors or experiencing other musicians as insensitive to them. They were able to perceive how their personal tensions affected the way they made music together. This resulted in their acquisition of greater nuanced musical expressiveness.

The relationship that evolved between these performers displayed aspects of the kind of rapport that Small (1998) referred to. The interpersonal themes were extracted through the filter of their family histories in music. For the music therapist, making this process conscious opened up new possibilities for connection in D2 in terms of musical collaboration. From a group therapy viewpoint as Yalom (1971) observed, the group allowed them to experience a corrective recapitulation of their primary family group. It seemed to create a new kind of musical family, one in which each could be safely heard and listened to.

Dimension Three: Connecting to the Audience

The first two dimensions are present in most music therapy experiences. Connecting deeply to music is a desired outcome for therapists who work with music as an expressive modality. Optimizing the musical connection between members of a group playing music together is a highly desirable social and communication outcome. Though groups may also serve as audiences for each other on occasions, the presence of an audience whose role is distinct from that of the performing musicians, is a phenomenon unique to the performance environment and one that can alter the experience of being in the first two dimensions. Evaluating how an audience alters the experience in dimensions one and two offers additional information to the clinician. It can enhance or disrupt connections to the music within or between performing musicians. Self-consciousness, anxiety or exhibitionistic tendencies may psychically protrude resulting in feelings ranging from constant inhibition and distraction to feeling the need to show off. The mere presence of an audience alters the experience. From a clinical assessment standpoint, the phenomenon of playing music in front of others allows the therapist to understand the degree of health or psychopathology that may exist. What patterns of behavior does the performer exhibit and what do they mean?

The audience effect may vary for the same person from each audience to each performance. Such variability may be an accurate reflection of certain variables like the room acoustics, the quality of the other musicians, ambient noise, difficulty of the music and the degree of preparation, or the health of the performer that day. However, certain people display patterns of behavior in front of an audience that seem less dependent on the variability of the performance environment and more the result of the internal state of the performer. This has been discussed in the music therapy literature in terms of developing working strategies to deal with performance anxiety in musicians (Berger, 1999; Montello, 1989; Montello, Coons, & Kantor, 1990). What has not been discussed in the literature is the significance of this phenomenon for people who may have been too frightened to perform and who were not working musicians. The work described herein is designed to address performance psychopathology for all people. Some may have been interested in performing but for many reasons, may not have had a chance to do so. For others, the experience was brief and buried in their past. For others still, mental illness may have terminated their careers prematurely. In each instance, discerning what performing before an audience means to the individual is the critical link. For the music therapist the salient concern is to understand how that individual operates in front of an audience.

For example one performer in the 2004 research group did not want to see the audience because it made her feel self-conscious "She performed best when she felt enveloped by the music and darkened by house lights. 'I don't want the audience to be known to me. They're not hidden. I want them hidden, no lights and not there'" (Jampel, 2007, pp. 125-6). The significance of this experience may lie in her relationship to her mother who at that time suffered from

Alzheimer's disease. Both of them were professional singers but her abusive mother constantly derided her and her voice saying that both were worthless. She reported that she tried to freeze these thoughts out of her mind when performing before an audience. She also said that the songs that she sang were songs that her mother loved and in this way, she was trying to reconcile with her ailing mother by performing her songs (Borczon, Jampel & Langdon, 2010).

Creating reconstructive experiences for the performer in the presence of an audience was a major group strategy. People not only learned about the importance of thorough preparation but perhaps more importantly, discussed the reasons why they had a hard time being on time and consistent in their work habits. These behaviors were examined from a psychodynamic viewpoint in which avoidance and withdrawal were possible signs of internal distress.

Practical skills were shared such as cuing each other on stage during the music and learning to look at and feed off each other's energy during concerts. In this way the group members were able to incorporate ideas for warding off distractions, improving focusing and concentration and building confidence through repetition. The power of the third dimension was huge. The group learned how to become each other's best audience. The feeling of reciprocal interplay going from performer to audience and back again was palpable. When that feedback loop did not happen during live performances, the performers attempted to compensate for that by sitting together in the audience and shouting vocal support. They would hush the crowd by standing up and rotating 180 degrees while putting a finger to their lips in an attempt to hush the crowd when the ambient noise became too much. Often they would hunch forward in their seats, applaud instrumental solos, and stand up and cheer wildly for each other. Handshakes, high fives and hugs would greet the performer as they returned from the stage.

Applause was the single most telling evidence of success for each performer. As one performer commented, she felt devastated when she received only "pity claps" after one particular performance (Jampel, 2007, p. 132). The quality of the applause seemed to trigger memories of her need for approval from her family. "I never had approval. If I wasn't good nobody said I was good... What I learned from that is I can't judge by my feelings. I never had approval" (p. 140). The issue of being applauded and what it meant to each person was a frequent topic of conversation in the group. The idea of how popular culture focused on audience approval came up when one member discussed her fascination with how contestants were judged and treated on the TV show *American Idol*. She commented in particular about the "spectacle" made by the daughter of another member of the band whose appearance on the show seemed to play up her obvious symptoms of mental illness. By appearing confused and rambling in response to not being chosen to advance to the next round, the group saw this segment as a shameless effort to humiliate a mentally ill performer. Several members of the group felt this episode was scary and depressing. What was even worse was the seemingly insatiable appetite of the audience to revel in the degradation of the performers who did not make it. It was thought by some that this reflected negatively on the larger cultural attitude of ridiculing vulnerable or less talented performers.

Some performers were more susceptible to patterns of self-doubt and lack of focus. For them a pre-disposition to anxiety states seemed rooted in historic patterns of harsh parental criticism and abuse. The treatment process involved working with them to gain insight into understanding the reasons for these patterns, developing the capacity to self-reflect when they began to emerge, and in promoting a stronger sense of self-worth in order to be able to withstand the emergence of these feelings in the group and on stage. The goal was to establish more stable emotional conditions within the performer so as to promote more musical consistency in each performance. This involved the working through of ingrained patterns of self-doubt, diminished self-worth and self-inflicted failures.

The audience in the After Hours Club was comprised of community members and friends and family of consumers. This gathering was mostly a familiar and friendly crowd. It was the kind of atmosphere that encouraged people and often inspired members of the audience to want to perform themselves. Many times over the years, someone would spontaneously get up and want come on stage to sing or play. An open microphone segment in the After Hours Club promoted this phenomenon that frequently led to the recruitment of new members for the band. However, when the performers went on the road, the audience experience was quite different. The room, the sound and the audience presented new challenges. The more experienced performers offered tips to the less experienced ones on how to read the audience, feel out the house and alter your play list accordingly. After having played a particularly tough house in the Bronx in 2004, one member of the performance group remarked, "You know what that proves? When you want to do something, you can do it. I don't think any of us in this group can say, no I can't." (Jampel, 2007, p. 151).

Dimension Four: The Audience Within

Performers carry their own audiences around inside of them. These are our music teachers and mentors, as well as our critics, juries and doubting parents. This audience is unseen to others but present in the mind of the performer. They may be heard intoning words of admonishment, advice, warning or support. These internal presences can serve a deepening, connective function. A face from the past associated to a particular song can evoke an authentic emotional memory. When produced intentionally it can provide an emotional basis for the music much in the same way an actor does in preparing for a role. It may also happen without design as when in the musical moment an internal association occurs that transports the artist to a particular image, feeling or place.

Performance however can also manufacture the presence of ghosts who hover in the shadows of artistic insecurity. When feeling the need to be perfect in order to gain special recognition or by getting caught up in feelings of competition with other performers, inner voices may be stirring underneath. These stirrings may be exhibited as anxiety, feeling distracted or as the fretful anticipation of a certain musical passage. Without understanding the source or origin of these issues, the performer may experience others as potential critics or rivals. Performance failures can be seen as self-fulfilling: the fear of the noxious element ends up interrupting the moment of musical flow that then leads to the undesirable outcome that was feared all along.

The capacity to harness the inner audience effectively is an evaluative landmark of creative health. Musical personalities that successfully internalize the presence of one's formative influences tend to display the capacity to bring them forth and integrate them into new and original experiences. Even when disturbances emerge, these performers can rapidly identify the triggers, know the origins of what is happening to them and through accurate self-observation, move through this process. However, when the internal presences are pathological, one sees an increase in the frequency and severity of these disturbances. The internal presences feel like an inner audience sitting in continual judgment of the performer. At its most extreme point, the performer is either frozen in a state of dreaded terror, unable to function, or overcome by a feeling that the inner audience has crept out of its internal sanctum and now sits in place of the real one. In such instances, this process may not be a part of the performers conscious awareness. This often results in the occurrence of distortions and projections.

The reality is that many performers' lives are stressful. Harsh critics not only exist but they abound. The music education system of much of Western music promotes performance pathology through rigid and highly inflexible standards of perfection and stressful competition. For the mentally ill musician whose inner life has been torn by unusually high levels of psychological trauma and cognitive difficulties, the toll is even worse. For people who are psychotic, the inner audience is real and alive. Helping them to focus on performing for the outer audience is the goal. Seven out of the fifteen members experienced visual or auditory hallucinations at some point during the duration of the group. Remarkably, making and performing music seemed almost invariably to provide them with relief from their psychosis. They each had the capacity to draw upon when healthy, a range of past memories and feelings that provided meaning to the act of performing.

Messages must be deciphered from the members of the inner audience to the performer. Who are they and what are they saying? This requires restoring some balance to the person's self-image when these inner states become unduly punishing. The leverage of the group is especially important here. It is more persuasive when eight people tell you that your singing is lovely than it is when there is only one. The structuring of feedback loops can break the cycle of self-deprecation. Conversely, when narcissistic features emerge that overvalue or hoard the spot light, a balanced group perspective may be more powerful than an individual therapeutic framework. Overall however, the likelihood of success for the narcissistic personality in music performance psychotherapy is more problematic due to the pathological social defensive structure of this personality type.

Through the experience of leading the Music Therapy Performance Group for two and a half years, the concept of the fourth dimension went through a process of expansion and clarification. I have come to appreciate the significance of the inner audience for assessment, treatment and evaluation. It seemed clear even six years ago that the inner audiences were present. The process and significance of how they were internalized and then projected onto the music performance experience was not then fully articulated or understood. The dualistic nature of performance with its inner and outer audiences has also evolved. I now believe that pathology in the other dimensions is most often the result of D4 pathology. One can liken it to the process of forming healthy/pathological attachments through the connection that develops in the musical relationship. Disturbances can impair the performer's ability to connect within themselves to music (D1), between others (D2), with the audience (D3), and resulting in a diminished sense of satisfaction (D5).

Dimension Five: The Totality of Experience

The fifth dimension (D5) has also undergone revisions with the passage of time. In 2004 several participants reported their performance experiences in terms of peak, altered or spiritual states (Jampel, 2007, pp.151-3). This author speculated that the experience of exultation occurred when all four dimensions optimally came together. This produced a heightened state of consciousness. Yet the reports also indicated that each performance experience differed, sometimes radically so. Disappointment or frustration occurring in one or more of the dimensions resulted in a range of feeling outcomes. For example, one person felt that her two most recent performances (in 2004) were "like a flip-flop. The best audience and the best time, the best everything that I had was up in the Bronx and then I can flip it around and say it was the worst" (p.130). She described her second performance like an "elevator that never got off of the first floor" (p. 130).

It is only reasonable to expect that there will be a variety of reactions to a given performance. Although peak states in performance were reported with greater frequency than in other situations, the evaluation of this phenomenon should take into account all nuances of experience. What seemed like a great concert to one person might seem very different to another. Though the variations in reports of felt experience were clearly present in the original research study, the discussion of the fifth dimension did not sufficiently factor into it the richness and diversity of how the performer felt about the performance and how useful this is as an evaluative device.

As I now understand it, the experience of the fifth dimension represents the totality of the performer's reactions. This includes the feelings of inner connection to the music, how that individual felt about the way the musicians played together, how they experienced the audience, and the felt presence of the inner audience. Together this might or might not rise to the level of a spiritual or peak moment but more importantly, it describes the precise contours of the feelings, concerns and thoughts of the performer. Here is how one group member described her recent performance experience, "It's a high when everything hits with the audience and you're clicking with the musicians and you're true to yourself. I can't think of any better expression. It's the totality of who you are" (Personal Communication, May 2008).

This dimension is based on the report of the individual and it is entirely subjective. If it seems exaggerated, distorted or extreme in its account of what happened as compared to the reports of other participants, the clinician's evaluation should consider the possibility of dimensional pathology. For instance, other group members who were present experienced the devastation that was felt by the performer who received "pity claps" after one particular performance, very differently. She felt that she sang miserably while others who were there (including this clinician) did not perceive this. One might postulate that the inner audience had been shouting negative comments. Some skewing of felt experience in performance is to be expected. However when the D5 experience of one person tends to reflect more consistent negativity than that reported by others, the likelihood is that pathological features exist in the person's performance personality profile. The evaluation process should follow the trail to the affected dimension(s) and explore the underlying causes. In the group, the music therapist should facilitate awareness of these issues, encourage feedback from others, and try to promote a greater sense of accuracy in the person's capacity for healthy self-regulation. The MTPG provided a supportive environment where feedback from other musicians who also frequently experienced musical insecurities, helped to balance out and even to outweigh those negative self-images. The reports of the D5 experience over time, reflected more consistently satisfying performance experiences.

Case Study

In May 2008, I conducted interviews with fifteen people who had received music therapy services in performance. I hoped to find out why they performed, what it meant to them, and what drawbacks they experienced in doing it. The following case study was partially drawn from one of these interviews.

Trisha was 63 years old at the time of the interview, an Italian-American female, divorced without children who had come into treatment in 1995 complaining of having experienced episodes of panic that prohibited her from working as a mid-level executive in a large corporation. She had recently divorced an Egyptian man who she reported was violent and abusive toward her throughout their marriage of eight years. Her early history revealed multiple episodes of abuse starting with her mother and brother who were reported to have been physically and emotional abusive toward her throughout her childhood and adolescence. Trisha was a trained dancer who also had participated as a member of several choirs throughout her life. Her younger sister was musical and according to Trisha, received all the praise and attention not because of her talents but because of her beauty. Neither parent was trained musically but her father was an opera lover and was the more loving and understanding parent. Trisha experienced her mother's attitude toward her interest in music as negative and discouraging. Her psychiatric diagnoses were Panic Disorder and Post Traumatic Stress Disorder.

Trisha has been involved with the band as a singer since 1997. Her ongoing difficulty in working cooperatively with the other members of the band, led to the idea that she could benefit from individual music therapy where she could focus on her own music. It was there that she began to compose her own songs and since then, has composed more than twenty songs that she has both performed and recorded. Songwriting offered a structure, which as DeNora (2000) observed, helped her to find a container for traumatic memories. Her music became an outlet for putting the past behind her and finding a way of moving ahead with her life. Her voice was another focus in individual music therapy. It was the target for reducing stress, improving her focusing and concentration, and providing greater emotional expressivity through changes in her vocal breathing technique.

Despite progress in these areas, she continued to labor in the band with difficulties in working with other musicians. She often erupted into arguments that spilled over into personal disputes that left her feeling victimized and abused. In the music, she had a tendency to race in front of the melody, sing out of turn, and find difficulty in maintaining accurate pitch. These disturbances in the music were closely connected to the interpersonal problems that she experienced. What became apparent was that her pitch and rhythm issues grew out of her problems with other people. Interventions strategies developed in band rehearsals to address her pitch and timing issues focused on promoting movement while singing so that she could both feel the music as a singer and as a dancer. This helped to a degree but recurrent personality issues continued to interfere. Despite efforts to address the destructive dynamics between her and other band members in rehearsals and in individual meetings, it became evident that too much time was being taken away from the music and from other band members without sufficient resolution of her interactional dysfunction.

After her entrance into the Music Therapy Performance Research Group, her psychotherapy needs could be attended to without time being taken away from rehearsals. No longer were individuals pitted against each other in proportioning time based on the need to get ready for the next performance. Now rehearsal schedules could be more closely adhered to. Once this inherent conflict was resolved, the level of tension between Trisha and the rest of the band eased considerably.

Dimensional Evaluation of Trisha

Trisha's performance experience was characterized by her harsh self-criticism that she reported in the following manner "I'll believe the negative before I believe the positive. I'm very self-critical. I have my mother's voice in my head all the time" (Personal Communication, May, 2008). From a dimensional viewpoint, one can look at this as D4 pathology. The intrusive, self-critical nature of her introjected maternal voice developed into a punishing musical super-ego. She was her own worst critic who often revoked permission to herself to complete a performance when she felt that she did not meet up to her own punishing standards of perfection "My self-consciousness takes over… which is extremely painful and embarrassing." This anxiety spilled over when she sang in front of an audience. "I am self-conscious in front of an audience. I always have been. That's why I always sang in groups." The anxiety of performing before an audience can be likened to singing in front of a room full of critics. She projected her D4 self-expectations onto others. The D3 audience effect can be seen as a pathological projection of her self-criticism. When activated in this way, Trisha's projections extended to her experience of working with other musicians in D2. Her relationship to the 53 year-old group member was sibling-like as he was perceived as an abusive brother. Her relationship with other females was characterized by anger and jealousy as was the case with the 78 year-old member whose deep and husky female contralto voice received much praise from others. Trisha often felt that she was given preferential treatment in rehearsals and said this in reference to her "I have had a very hard time in the band. We had women who had powerful lower voices than I did. I was very high and lyrical. You know I was criticized for whatever." Ironically, the 78 year-old was an abuse survivor herself and had always been belittled by her musician father for her voice that he regarded as unfeminine. Praise was hardy something she assumed was forthcoming.

It took Trisha time to place me in her constellation of family characters. At first, she experienced me as favoring other people. She also said that I never had a complimentary word for her music. It was only later on that she experienced me as being more caring and appreciative of her. I would say that her transference could be described as similar to her relationship with her mother. I seemed unappreciative of her talents and often I was perceived as highly preferential to others. Eventually I defied easy description, as a more healthy D2 connection became her dominant perception of other performers, myself included. But first we had to work on her overtly critical D4 issues in order to restore her capacity for self-worth. Over time she was able to balance out how she saw herself and consequently, how she saw herself in relationship to others.

All of this pathology on D4, D3, and D2 took a tremendous toll on her D1 inner connection to the music. "I have been judged and have been criticized but no one is worse at criticizing me than myself. I have had to let go of that." As she did so, she began to experience music differently. It was like she could hear it better, reproduce pitch more accurately, feel the rhythm and tempo better, remember the words and connect more deeply. She stopped fighting in rehearsals, moved more fluidly and became more expressive. The reaction of audiences was telling. She experienced recognition even admiration for her performances and as compliments came her way, "finally one day I decided to believe it."

Her D1 pathology looked like never being sure which song was right for her. Often she felt the register was too low or too high for her voice, the key was wrong, it needed to be faster or slower and finally, she just wanted to move to another song. With the breakthrough in her D4 and D1 pathology, new channels were opened to her. She re-examined her experience of playing with other musicians in the MTPG and began using the group as an opportunity to become more comfortable with her D1 connection to the music. Her vocal duets solidified her D2 connection. The group became a new kind of supportive audience that encouraged a healthier D3 connection. One day she received a standing ovation in the After Hours Club "that put me over. I crossed over from timidity, if that is a word, over to the other side." When asked about the importance of playing in front of an audience she responded:

> Yes, I got a lot of confidence by trusting in myself and the feedback that I have gotten. I'm really appreciative. I have never taken it for granted. Each time I perform it's a new day, a new performance that I don't take for granted. I will achieve what I want to achieve.

No longer did she transfer sibling issues onto other performers but now she could take in what they had to offer without seeming to flinch or duck as she had before. "They have come to accept me. We don't have that friction anymore." The improved regulation of projections from D4 onto D2 and D3 allowed her to concentrate better and focus on D1, D2 and D3 without as many intrusions. The synergistic effect of all levels flowing together produced a sense of lift and well being in her. She was able to balance out her perceptions of herself with what others thought of her. "I just balanced myself because if I'm satisfied with a performance, it doesn't matter what someone else thinks."

As her self-critic eased up, Trisha's sense of accomplishment came through more often. "Sometimes I'll be very good and sometimes I won't (laughs) but that doesn't take away from my expressing myself." The experience of performing music on D5 portrayed a person who had a greater degree of personality integration and realistic observing ego. This was evident when she said, "I feel like a whole person. I feel very cleared to all negativity. I may not always think that I performed my best but that's just a performance. It's not taking my voice away and forgetting who I am."

The quality of being more fully in each moment reflected a person who had managed to free herself of her inner tormenter. "However I could express myself at that moment is the best that I could do at that moment and there always is another moment." This fluid state of being characterizes a healthier personality that has opened more her optimal creative *flow* (Csikszentmihalyi, 1990).

It was not surprising that her D5 experience summed up how performance offered her transcendence.

The spontaneity of being in the moment and feeding from the audience and seeing their faces light up. I still could look up or down, close my eyes. I could look at the audience and feed from the audience and the other musicians. If they are playing well and if they are in sync with me and I am in sync with them and you're feeding off each other, it's a tremendous experience.

In this statement, Trisha summed up the essence of how it feels when all the dimensions are aligned for optimal *flow*. The "feeding" from the audience, the "in sync" feeling when you are locked in with other musicians, and of being 'true to yourself' in the sense of connecting with your inner music and to your inner audience.

For Trisha, performance brought out her playful side. It allowed her to dress up in costumes, put on make-up, and use her talents as a dancer and actor. She found new ways to interpret the emotions of the characters in her songs. She composed music that employed her talents for poetry and writing. And perhaps most significantly, she found a new family that recognized and appreciated her talents.

Conclusion: Brothers and Sisters in Performance

The connections established in the MTPG produced a strong bond. A feeling of shared experience occurred that promoted a culture of belonging to some bigger entity than oneself. The traditions of performers, their customs, common language, their encounters and experiences, the sense of being part of a larger community of artists, was all felt. The relationships that evolved in the group explored, affirmed, and celebrated the empowerment of the performer. The members came to embrace performance as the domain of the committed not just the gifted.

Ansdell (2004) describes the experience of *communitas* (first used by the anthropologist Victor Turner and discussed by Even Ruud) in referring to the feeling of membership promoted by performance. Aigen (2005) uses this term to denote the camaraderie that develops between members of a band through the ritual of playing music together. He discusses the "liminal" or timeless qualities of performing as "losing oneself in the experience, leaving behind symbols and practices of previous positions, ambiguities, perceived danger, the absence of roles and a transcending of previously defined borders" (pp. 91-92).

Despite the strong presence of camaraderie that developed, inevitably individuals continued to struggle. The liminal aspect of the performance process can be deceptive. It can overshadow the personal experience of one performer who may not feel the same glow that is shared by the rest of the group. A therapist should not assume that all moments even the peak ones, are being felt in the same way by each person. What often develops is a special understanding between performers. The experience comes to feel more like a band of performing brothers and sisters in the way that people come to feel about having gone through a lot together even if it was not always easy. One difference between this group and their own families of origin however was in the support and encouragement that they found that was often missing there.

References

Aigen, K. (2004). Conversations on creating community: Performance as music therapy in New York City. In M. Pavlicevic & G. Ansdell (Eds.), *Community music therapy* (pp. 186-213). Philadelphia: Jessica Kingsley Publishers.

Ansdell, G. (2005). Being who you aren't; doing what you can't: Community music therapy & the paradoxes of performance. *Voices: A World Forum for Music Therapy*, North America, 5, Nov. 2005. Retrieved January 5, 2011, from

Aigen, K. (2005). *Playin' in the Band: A qualitative study of popular music styles as clinical improvisation*. Gilsum, NH: Barcelona Publishers.

Ansdell, G. (2004). Rethinking music and community theoretical perspectives in support of community music therapy. In M. Pavlicevic & G. Ansdell (Eds.), *Community music therapy* (pp. 65-90). Philadelphia: Jessica Kingsley Publishers.

Berger, D. S. (1999). *Toward the Zen of performance: Music improvisation therapy for the development of self-confidence in the performer*. St. Louis: MMB Music, Inc.

Borczon, R. M., Jampel, P. & Langdon, G.S. (2010). Music therapy with adult survivors of trauma. In Stewart, K. (Ed.) *Music therapy & trauma: Bridging theory and clinical practice*, New York: Satchnote Press.

Csikszentnihalyi, M. (1990). *The psychology of optimal experience*. New York: Harper and Row.

DeNora, T. (2000). *Music in everyday life*. Cambridge: Cambridge University Press.

Inoue, S. (2007). A study of Japanese concepts of community. *Voices: A World Forum for Music Therapy*, North America, 7, Jul. 2007. Retrieved January 7, 2011, from:

Jampel, P. (2007). *Performance in music therapy with mentally ill adults*. Dissertations Abstracts International. (UMI, Order #3235696).

Maslow, A. (1971). *Farther reaches of human nature*. New York: The Viking Press.

Montello, L. (1989). *Utilizing music therapy as a mode of treatment for the performance stress of professional musicians*. Doctoral dissertation, New York University. (UMI Order No. 9004310).

Montello, L. Coons, E.E., & Kantor, J. (1990). The use of group therapy as a treatment for musical performance stress. *Medical Problems in Performance Art*, March (5), 49-57.

Nzewi, M.(2006). African music creativity and performance: The science of the sound. *Voices: A World Forum for Music Therapy*, North America, 6, Mar. 2006. Retrieved January 6, 2011, from:

O'Grady, L.(2008). The role of performance in music-making: An interview with Jon Hawkes. *Voices: A World Forum for Music Therapy*, North America, 8, Jul. 2008. Retrieved January 7, 2011, from:

Oosthuizen, H., Fouché, S., Torrance, K. (2007). Collaborative work: Negotiations between music therapists and community musicians in the development of a South African community music therapy project. *Voices: A World Forum for Music Therapy*, North America, 7, Nov. 2007. Retrieved January 7, 2011, from:

Small, C. (1998). *Musicking*. Hanover, New Hampshire: Wesleyan University Press.

Stige, B. (2002). The relentless roots of community music therapy. *Voices: A World Forum for Music Therapy*, North America, 2, Nov. 2002. Retrieved January 6, 2011, from:

Stige, B., Ansdell, G., Elefant, C., & Pavlicevic, M. (2010). *Where music helps: Community music therapy in action and reflection*. Ashgate Publishing Co.

Turry, A.(2005). Music psychotherapy and community music therapy: Questions and considerations. *Voices: A World Forum for Music Therapy*, North America, 5, Mar. 2005. Retrieved January 5, 2011, from:

Wood, S (2006).. "The matrix": A model of community music therapy processes. *Voices: A World Forum for Music Therapy*, North America, 6, Nov. 2006. Retrieved January 6, 2011, from:

Yalom, I. (1970). *The theory and practice of group psychotherapy*. New York: Basic Books.

©VOICES:A World Forum for Music Therapy.

CHAPTER 12

MUSIC THERAPY IN HOSPICE AND PALLIATIVE CARE

Joey Walker
Mary Adamek

CHAPTER OUTLINE

HOSPICE AND PALLIATIVE CARE: WHAT ARE THEY?
WHO BENEFITS FROM HOSPICE OR PALLIATIVE CARE?
MUSIC THERAPIST AS PART OF THE HOSPICE TEAM
ISSUES AT END-OF-LIFE, TYPICALLY ADDRESSED BY MUSIC THERAPISTS
  Physical Issues
  Psychosocial Issues
MUSIC THERAPY GOALS AND INTERVENTIONS COMMONLY USED IN HOSPICE AND PALLIATIVE CARE
  Music Therapy to Alleviate Physical Symptoms
  Music Therapy for Psychosocial Support

Mike, a 60-year-old male, had recently moved into a long-term care facility because of complications with his cancer. He was no longer able to care for himself at home. The home health aide had reported that Mike did not want to bathe, get dressed, or do his other activities of daily living. When the music therapist visited late one morning, Mike was lying in bed, unshaven, and he had not put his false teeth in his mouth. He was unkempt and not dressed. He agreed to a music therapy session, and 45 minutes later he was energetic and reminiscing about good times when he and his significant other would dance together in numerous small towns around the area. He asked the music therapist to return the same time the next week. Upon returning the next week, Mike was clean, dressed, and lying on the top of his bed. Soon his significant other entered the room and slid onto the bed with Mike. They were able to share a songbook together, sing, talk about dancing, and express their love for each other.

HOSPICE AND PALLIATIVE CARE: WHAT ARE THEY?

The terms *hospice* and *palliative care* refer to a philosophy of care for people at end-of-life. This **team-oriented approach** provides compassionate **end-of-life care** to enhance comfort and improve quality of life for individuals who have **terminal illness** and for their families. The goals are to prevent suffering, relieve pain, and optimize each person's functioning. The individual's decisions about care are central to the hospice and palliative care philosophy, and the team is guided by the wishes of the patient and family.

People are eligible for hospice treatment when their death is anticipated in six months or less. Palliative care can be provided for persons who are dealing with terminal illness whether or not they have a six-month prognosis. The terms *hospice* and *palliative care* are closely related. Both are concerned with providing relief but not cure, and in some parts of the world the terms are used interchangeably. The term *hospice* will be used throughout this chapter to indicate a model of care to improve quality of life for people with life-limiting illness and for their families.

WHO BENEFITS FROM HOSPICE OR PALLIATIVE CARE?

Adults and children with terminal illness can be admitted to hospice care. The largest number of patients have a cancer diagnosis, while others are admitted with heart, lung, or **neuromuscular disease; Alzheimer's/dementia;** organ failure; HIV/AIDS; or other disorders. Adults and children may have developmental delays concomitant with their medical diagnosis. Children have different physical and psychosocial needs than adults, based on their ages and developmental levels. The family is also a recipient of hospice care while the patient is dying and during **bereavement** after the family member's death. The interdisciplinary team addresses the unique needs of each family unit throughout the dying and bereavement process.

MUSIC THERAPIST AS PART OF THE HOSPICE TEAM

The goal of hospice is to care for each person in a holistic manner. This requires an interdisciplinary team to address patients' physical, psychological, spiritual, and social needs. The interdisciplinary team plans coordinated care, holds regular team meetings, and continues ongoing communication to ensure that goals are met and frequently reassessed. The team includes the primary physician, hospice physician, nurse, social worker, chaplain, home health aide, bereavement counselor, and volunteers. Additional team members may include a music, occupational, or physical therapist; psychologist; pharmacist; and nutritionist, among others. Patients and families are considered part of the team and are able to direct their desired care by communicating specific needs to the rest of the team (National Hospice & Palliative Care Organization, 2008).

Music therapists utilize comprehensive skills to observe, report, document, and provide effective interventions. A music therapist may provide treatment for physical, emotional, spiritual, cognitive, or social needs. Viewing the whole person as the interaction of mind, body, and spirit, the music therapist has a unique place among other professionals on the team. Other team members may concentrate their efforts mostly in one area of expertise, such as with physical or spiritual needs. The music therapist may be able to offer insight for team members concerning the multidimensional needs of each patient.

Hospices are becoming more aware of the benefits of providing music therapy services. Music therapists provide support in a noninvasive, cost-effective approach. Music therapy in hospice care is one of the fastest growing areas in the field of music therapy, with the creation of many new employment opportunities in the last few years (American Music Therapy Association, 2007).

ISSUES AT END-OF-LIFE, TYPICALLY ADDRESSED BY MUSIC THERAPISTS

The needs of hospice patients and families vary greatly, and these needs may rapidly change from day to day, hour to hour, and within a single session (Krout, 2000). Therefore, the music therapist will provide services that concentrate on physical, **psychosocial**, spiritual, and bereavement needs of the moment for each session. For example, the music therapist may have concentrated efforts on a spiritual issue during a past session, and in a present situation may focus on pain control. A session in the future could consist of using music to stimulate memories and life review, or perhaps any combination may take place within a single session. Ongoing assessment is critical to ensure that the patient and family are receiving the care that they desire.

Physical Issues

Pain management. Pain management is a primary focus for the hospice team. Although not every patient in hospice care has pain management needs, pain is still the most common symptom experienced by hospice patients (Kastenbaum, 2001). Patients sometimes improve in hospice care because the team is able to find effective ways of treating the **total pain** of each patient. Cicely Saunders, who is considered the modern founder of hospice, created this concept of "total pain" in order to ensure that psychological, emotional, social, spiritual, as well as physical pain of patients and their loved ones is included in treatment (Hilliard, 2005). This concentration on the whole person interfaces easily with music therapy, as the music therapist can simultaneously address several goal areas with specific interventions.

Hospice care takes into account the desired level of pain management for each patient and family. Because pain is subjective and complex, each person's experience differs from another. For a variety of reasons, people also have different levels of pain that they will tolerate at any given time. One patient may want to be alert and thinking clearly when making legal or financial decisions, while another patient may want to be able to be fully awake to visit with a long unseen family member. Some families do not want to use pain medications for their loved one due to fears of addiction or sedation, and other patients are unable to use medications because of certain symptoms, disease process, allergic reaction, or other undesirable side effects. Some patients may not receive pain medication in a timely fashion, if at all. For others, generational, spiritual, or cultural considerations or stoicism may contribute to reluctance in admitting that pain is actually present. If untreated, unrelieved pain may lead to the following:

- Fatigue
- Stress
- Nausea
- Loss of appetite
- Isolation
- Anxiety
- Difficulties with daily activities
- Disrupted sleep patterns
- Depression
- Relationship difficulties
- Anger
- Thoughts of suicide

(National Foundation for the Treatment of Pain, 2008)

In addition to the problems listed above, patients may use much of their energy to deal with unrelieved pain. They may have little energy remaining to take care of other essential end-of-life issues like emotional problems or spiritual pain. Conversely, Trauger-Querry and Haghighi (1999) discuss the fact that treatment for pain can be resistant if psychosocial, emotional, or spiritual issues are disregarded.

Pain assessment. It can be challenging for adults to admit, describe, and discuss issues of pain. Children's expression and understanding of pain is compounded by their level of development. Instead of verbalizing about pain, children may exhibit behavioral distress, which may involve changes in behavior, sleep, or eating patterns; becoming withdrawn; decreased physical activity; increased irritability; or an increased need to seek comfort (Barrickman, 1989). Preschool children may not be able to verbally describe their pain or anxiety and may act out behaviorally, for example, by screaming, hitting, or having a tantrum. Older children may withdraw and become quiet as a means of control and may not verbally express pain because of fear of receiving painful procedures or treatment. Adolescents may have difficulty communicating their needs in general, and admitting to pain may keep them from spending time with peers or may curb their independence. Within a normalized musical environment, children may demonstrate more congruence with feeling and verbalization and may also express themselves on an emotional level more easily with music as the stimulus (Ghetti & Walker, in press).

Chapter 10 described a number of **assessment tools** that may be useful in a hospice or palliative care setting. Because of the nature of hospice care, the choice of a suitable assessment tool should take into account a number of factors: the functional level of the patient, an assessment that is as nonintrusive as possible, and an assessment that can be completed quickly, given what is sometimes a rapid change in patient status.

Music therapists have access to a variety of formal assessment tools with regard to pain and discomfort. Pain assessment tools such as **Numeric Rating Scale (NRS)** and Faces are recommended for use by nonnursing team members (Mills-Groen, 2007). The patients rate their perceived pain by choosing a number from the scale (NRS) or a picture/line drawing (Faces) of a face of a person experiencing different levels of pain severity. These assessment tools are easy to use and take little time to determine the patient's pain intensity at the moment.

A music therapist must be able to continuously observe in a less formal manner these possible indicators of pain:

- Vocal and verbal complaints
- Rubbing, holding a body part
- Bracing
- Facial grimacing/winces
- Furrowed brow, frown
- Irritability
- Anxiety
- Restlessness
- Physical repetitive movements
- Repeating words or phrases
- Change in behavior from the norm

(Cohen-Mansfield & Creedon, 2002;
Maue-Johnson & Tanguay, 2006;
Warden, Hurley, & Volicer, 2003)

Lower functioning patients may be unable to verbally express their pain and its symptoms; therefore, music therapists carefully observe for the indications listed above and monitor changes in behavior. Often the anxious or restless behavior of a patient in a long-term care facility may appear to be related to a medical condition such as dementia; however, it may be caused by unresolved pain. Other indicators of unaddressed pain in lower functioning patients may include:

- Eyes tightly closed
- Tense muscles, clenched fists
- Increased pacing
- Changes in sleep patterns
- Wanting to exit home or facility
- Increased agitation
- Pulling away or hitting when touched
- Decreased appetite

(Warden et al., 2003)

Symptom management. Symptom management is an integral part of hospice care. Similar to pain management, symptoms may not be eliminated but managed at a level desired by the patient or family.

Dyspnea. **Dyspnea** is shortness of breath that occurs in patients with a variety of life-limiting conditions, but commonly occurs in **chronic obstructive pulmonary disorder (COPD)** and with some cancers. This feeling of not being able to get enough breath or suffocating causes anxiety, which may lead to increased dyspnea which subsequently may lead to increased anxiety in an unlimited cycle (Hilliard, 2005).

Agitation. Agitation can be a significant problem for patients and families. While there are many causes of agitation such as pain or disease process, agitation can lead to safety issues and be distressing for staff or caregivers.

Sleep difficulties. Patients may have problems regulating their sleep cycles. They might sleep during the day, making it difficult to fall asleep at night. Sleep problems can contribute to increase in anxiety and agitation.

Restlessness. Near the end-of-life when multiple body systems are shutting down, some patients (even those who have previously been calm) may experience terminal restlessness. This may include anxiety, thrashing or agitation, palpitations, shortness of breath, insomnia, pain, moaning, yelling, involuntary muscle twitching or jerks, fidgeting, or tossing and turning (Hospice Patient Alliance, 2008). This can be upsetting for the patient, family, and staff who are also coping with other end-of-life issues.

Psychosocial Issues

In hospice care, patients also face end-of-life psychosocial and spiritual issues. As noted in Chapter 10, medical conditions greatly impact a person's psychological, social, and emotional functioning. This impact on quality of life, according to Krout (2000) and Hilliard (2003), can be improved through music therapy psychosocial support. When patients reach a desired or tolerable level of physical comfort, they may have the basic energy or need for emotional support and comfort. Due to the sizeable amount of end-of-life psychosocial needs, concentration will be given to coping with illness and loss through the following areas: (1) anxiety reduction, (2) emotional support, (3) autonomy and control, (4) reduction of isolation, and (5) family cohesion.

Anxiety. Anxiety related to dyspnea was previously discussed; however, anxiety can also be caused by psychological, social, or spiritual factors. A terminal prognosis can create fear of the unknown and manifest in tension and anxiety (Hilliard, 2005). Patients and families may want to know why or how someone will die, if it will be painful, and when it will occur, etc. Family members may not agree on the plan of care (e.g., withholding aggressive treatment or measures given for comfort only), where to spend the last part of life, or merely how to care for the patient. This may compound stress and tension in an already unstable situation and cause more anxiety for the patient as well. Family members may have additional anxiety thinking about and dealing with unknown issues.

Emotional needs related to losses and spirituality. People experience many losses throughout the dying process. Patients may have the role loss that may have helped define who they were. For example, a mother who was the constant caretaker of everyone in the family who can no longer fulfill that role, or a child who cannot attend school, may feel the loss of who he or she is. Patients may also physically appear different and may have lost home and family; they may need an outlet that helps them feel serene, one that can help them express who they are. Patients and families need a way to express their personal identity, search for meaning, understand their relationship to God or Higher Power, and complete unfinished business (Trauger-Querry & Haghighi, 1999).

Each patient guides the depth of emotional involvement for the team according to his or her own individual needs. Patients may need someone who will take time to listen and validate their feelings, and at times they may feel more comfortable expressing feelings to hospice staff instead of a family member. Patients may not want to upset someone in their family or may feel that they shouldn't be feeling a certain way, or that it is wrong to feel an emotion that might be considered negative. It is also possible that they may be unable to verbally express emotions or they may not know what they are feeling.

In general, people of all ages and levels of functioning need help when coping with illness or loss. At end-of-life the problems may be magnified and may feel overwhelming. At times patients may experience a series of losses that appear to never end. Patients may lose independence with the inability to physically or mentally do what they desire, may have chronic and acute pain, may have moved to a different environment and are mourning the loss of a familiar place, may have to adjust to new caregivers and the stress that accompanies loss of privacy or dignity, may have financial concerns or feel like they are a burden, or may be separated from friends or loved ones. In addition to adjusting to a medical condition and their own mortality, patients may have these and other losses that make coping more difficult and contribute to emotional and spiritual pain.

Many people find a need for spiritual comfort at end-of life. If a hospice is **Medicare-certified,** it is required to provide a chaplain for spiritual support. However, the music therapist also provides comfort and additional spiritual support.

Autonomy and control. In some patients, feelings of helplessness and low self-esteem surface with the loss of normalcy and independence. Motivation to attempt straightforward daily responsibilities may decrease and depression may become more likely when people are unable to accomplish simple tasks. Patients who have developed dependency and helplessness may have difficulty making even a simple forced choice.

Isolation. Isolation is a common problem for many patients in hospice care. People may live alone at home or in long-term care facilities, or are in hospitals where it is more difficult for various people to visit. Many may be of an age where most of their friends and family have died. Others do not want people around to witness their decline. Some people avoid visiting friends at end-of-life, while other visitors do not know how to interact with someone who is withdrawn or may not communicate in a familiar way.

Socialization is an important component of quality of life (Hilliard, 2005). People have a need to feel as if they are accepted and belong. Integrating acceptance and belonging into musical interventions is typical and inherent in the musical process (see Chapter 3). When feasible and appropriate, group interactions take place within families, with friends, or with other residents of facilities. Making music together helps develop a sense of belonging, whether in a group situation or simply with the music therapist.

Family cohesion. When people gather at the end-of-life, there may be family members who have not seen each other since they were young. They may disagree on issues, and old patterns and resentments may reoccur. It can be a stressful time, with family members missing time from work and their own family responsibilities. Disagreement can be distressing for the patient as well as the family.

MUSIC THERAPY GOALS AND INTERVENTIONS COMMONLY USED IN HOSPICE AND PALLIATIVE CARE

The most widely used intervention, listening to live or recorded music, can reduce pain perception and anxiety; provide relaxation, comfort, and spiritual support;

and offer a means for life review and emotional expression. The music serves as a stimulus for active listening, which can offer a means for reminiscence, verbal discussions, and emotional expression. Music can also provide for a more passive means while creating a positive sensory environment, for enhancing relaxation, or for reducing anxiety and agitation (Krout, 2000).

According to Krout (2000), hospice and palliative care music therapists often use a combined treatment strategy to address multiple patients' needs and goals. Many interventions may be used in the treatment course or within a single session in order to provide individualized care; however, the techniques most often used are the following:

- Music listening
- Improvisation
- Singing
- Songwriting
- Music playing
- Song choice
- Music/imagery for relaxation—with progressive muscle relaxation and deep breathing
- Lyric analysis—music assisted cognitive reframing

(Hilliard, 2003; Krout, 2000; Mills-Groen, 2007)

Instrument playing, singing and *improvisation* are effective techniques for facilitating expression of emotions and improving communication with terminally ill patients (Krout, 2000). Many different kinds of instruments and styles of music can be utilized according to the preference of the patient. People of all ages and functioning levels are able to participate with adaptations prepared by the music therapist. Patients may play with great expression and musicality within a successful experience designed to foster creativity, enhance self-concept, and provide a means to the unconscious. Emotions not easily verbalized such as anger, fear, and existential concerns may be expressed unconsciously through improvisation (Krout, 2000).

The universal appeal and novelty of live music is an effective tool for distraction. *Listening to live or recorded music* may help patients focus their attention on something other than pain, **perseverative behavior**, painful procedures, daily cares, worry, or anxious thoughts. Distraction can be used with people of all ages and functioning levels; however, it works particularly well with infants and young children. Providing distraction with instruments of a **vibrotactile** nature may be effective for older patients with cognitive impairments. Utilizing a variety of easily adaptable and colorful vibrotactile instruments, which are both aurally and visually appealing, in combination with the skills of the music therapist allows for successful distraction.

Songwriting is a valuable technique for assisting patients with creative and emotional expression or self-awareness, enhancing self-esteem, providing validation, and creating a lasting gift for a loved one (Krout, 2000). A variety of songwriting techniques can be adapted for differences in functioning levels as well as age. Music therapists can design the music so that it is applicable to the situation and considers the musical preference of the patient as well.

Music-based cognitive reframing and *lyric analysis* use music to stimulate discussion about thoughts and feelings. Music listening, singing, songwriting, and song choice can all be used as a means for discussing the lyrics as they relate to a patient. Cognitive reframing refers to changing the way a person would view a situation by changing the way he or she would think about it.

Music can provide the structure for slow, deep breathing, and imagery may be added after *autogenic* or *progressive muscle relaxation* has occurred. Autogenic relaxation uses self-directed visual imagery like the repetition of a word, phrase, or feeling combined with body awareness. Progressive muscle relaxation involves slowly tensing and then relaxing different muscle groups within the body (Mayo Foundation for Medical Education and Research, 2008). The use of imagery in order to reduce anxiety or provide relaxation can be used with music assisted relaxation (MAR). However, in order to use the approaches of Guided Imagery and Music (GIM) or the Bonny Method of Guided Imagery and Music, one needs advanced training and certification. These methods also use music, imagery, and relaxation, but the goals of the practice are creativity, self-exploration, insight, and reorganization (Krout, 2000).

Music Therapy to Alleviate Physical Symptoms

Manage pain. Music therapy offers a comprehensive **nonpharmacological approach** for pain management, and there are many ways that music therapy interventions can be utilized to reduce pain perception. Because of the fragile interactive dimensions and complexity of total pain, music therapy as a multifaceted treatment is an effective modality for pain management. Standley (2000) compiled music research in medical treatment and generalized that music is most effective when:

- a patient experiences mild to moderate pain. As pain becomes severe, music is less effective.
- live music (as opposed to recorded music) is provided by a music therapist.
- it is patient preferred.

See Chapter 10 for descriptions of the following music therapy interventions associated with pain management:

- as a stimulus for **active focus** or **distraction**
- to facilitate a **relaxation response**
- as a **masking agent**
- as an **information agent**
- as a **positive environmental stimulus**

Jean was an elderly woman with dementia and severe arthritis who lived in a long-term care facility. Her days consisted of sitting in a reclining wheelchair with her eyes closed or lying in her bed. She sometimes would answer a closed question with "mm-huh" (yes) or a shake of her head meaning "no."

Because of her arthritis, Jean's hands had contracted so tightly that her fingernails were causing open sores in the palm of one of her hands. A hospice home health aide would soak and gently massage her hands over a period of time to help open them. The area could then be cleaned, medicated, and dressed. Even with pre-medication, this was a painful process; therefore, the music therapist would provide distraction, relaxation, deep breathing, and imagery to assist with pain control.

Jean would cry out and moan in pain during this process, so the music therapist provided live, slow, arpeggiated guitar accompaniment that matched the pitch of her moaning. The music therapist hummed the pitch of her moan and began gradually dropping the pitch as if in a sigh. Intermittent breathing in an audible manner by the music therapist (inhaling and exhaling slowly with music as the guide) as well as giving Jean cues to "keep breathing" gave her both a focal point and distraction from her pain. Jean began to follow the music therapist's drop in pitch with her moans as well as breathing more evenly and slowly. The music therapist continued to hum and sing, improvising and weaving images about Jean's farm into the music. Jean's grimace and furrowed brow disappeared and her face relaxed. Jean stopped moaning and listened as the music therapist sang about the sights, smells, sounds, and general feel of her farm on a warm, humid summer day. The nurse and the home health aide both reported that Jean had significantly less pain and discomfort when music therapy was provided during her dressing change. In addition, these caregivers also stated that they felt "much less stressed" during the procedure with support from music therapy.

This situation involved music therapy procedural support for acute pain. Music therapy provided the multifaceted approach that Jean needed to reduce her pain. The music therapist used the **iso-principle** (matching the patient's mood with the music) to provide sensory input, as well as cognitive strategies, breathing techniques, distraction, and a focal point in order to reduce Jean's pain perception.

Promote relaxation.

> Bud's health had been declining rapidly for the last year; he lay in bed and was unable to move his arms. He was recently diagnosed with **amyotrophic lateral sclerosis (ALS)**, was becoming weaker, and had increasing pain and difficulty swallowing foods and liquids. His anxiety was increasing due to dyspnea. He had played guitar in the past and still enjoyed watching movies and listening to music. The music therapist provided live vocal and guitar music as requested by Bud.

With the knowledge and skills of the therapist, the music was able to serve multiple purposes:

- As a focal point—Bud did not think about his pain or dyspnea as he concentrated on the lyrics and guitar.
- As a way of reducing anxiety through music assisted relaxation—As time progressed and Bud became more anxious, music was found to be the most comforting and effective means for relaxation. The music therapist was able to provide live sessions, recordings, and cognitive strategies with MAR to help Bud reduce his dyspnea and anxiety.
- As a stimulus for reminiscence—Bud was able to recall many happy times when he played the guitar. He was able to remember specific performances and how he felt at the time. This was a way of life review and validation of his life.
- As a means for spiritual support—Bud was unable to attend the services of his faith tradition, and he felt close to his Higher Power when his favorite spiritual music was provided.
- As a stimulus to improve social interaction—Bud's friends and family sometimes felt uncomfortable interacting with him. The music provided a vehicle for socialization. His grandchildren as well as his friends could all participate together and interact in a normalized way.
- As a means for control—Bud was able to make choices (e.g., fast or slow music, type of music, specific song, etc.) within an environment where he was slowly losing more and more control everyday.
- As a stimulus for finding meaning and purpose—Bud was able to teach the music therapist some advanced concepts for guitar. This helped give him purpose

and improved his self-concept as he was still able to help someone else despite his compromised health.

Bud's case illustrates the point that many patients in hospice care have more than one need. Bud had physical needs with his chronic pain and shortness of breath, as well as many psychosocial and spiritual needs. When a patient has more than one need, the music therapist must be able to assess quickly what need is most important to focus on at the time. Flexibility and the ability to change, modify, or create new plans at any time are vital when working in hospice care.

Using MAR at end-of-life can be an effective and powerful tool to relieve pain, reduce anxiety, and provide relaxation and a calm, soothing environment. However, level of functioning of the patient and disease progression need to be considered. Patients in the later stages of illness may not be able to participate in long relaxation and imagery sessions; therefore, shorter sessions may be advisable (O'Callaghan, 1996). Lower functioning patients may not be appropriate candidates for imagery but may be able to concentrate on breathing techniques paired with music. Patients with a history of emotional problems, abuse, low mental energy, or problems with concentration or reality may be better served by more passive music relaxation. MAR can be used with children when consideration is given to developmental level, goal of the approach, and type of intervention—sedative music listening, music facilitated deep breathing, music and imagery, or progressive muscle relaxation (Ghetti & Walker, in press).

Music can be tailored to calm, soothe, and orient people according to their specific requirements and musical preferences. Some lower functioning patients or those who have dementia may have increased agitation with daily cares such as bathing, dressing changes or other activities. Music therapists can co-treat and provide distraction and relaxation during procedures or daily cares to reduce agitative behaviors. A patient who is restless may be easily distracted or soothed by carefully administered musical stimuli. If a patient is observed to be restless, the music therapist can often reduce the likelihood of the patient becoming agitated. The music provides the same result as medication, but without the negative side effects that may accompany medications often administered to reduce agitation or restlessness.

Adjust sleep cycles. The music therapist can provide live or recorded music for people who have sleep problems. If patients sleep during the day, they are less likely to fall asleep easily at night, or remain asleep for a period of time. Therefore, stimulating music may be needed to keep the person awake during the day. Conversely, sedative music can be provided in order to help patients fall asleep. This can be a positive nonpharmacological strategy to assist patients who often do not want more medication. Taking less medication is less stressful for the patient and a cost saver for families and facilities. Music can also help mask unwanted noise from a hallway, a roommate, or medical equipment, while providing comfort.

Music Therapy for Psychosocial Support

Patients may be referred for music therapy services for an extensive range of psychological, emotional, and social issues including anxiety, depression, isolation, confusion, grief, impaired communication, ineffective coping, normalization, self-esteem, control, relationship/family problems, diversion, lack of insight, life review and reminiscence, disorientation, and motivation (Dileo & Dneaster, 2004; Krout, 2000; Maue-Johnson & Tanguay, 2006; Mills-Groen, 2007).

> *The music therapist enters the room of Dan, a 55-year-old nonresponsive male patient on the palliative care unit located within a hospital. Many family members are present; however, they are not interacting with each other. The television is on and they are sitting quietly around the perimeter of the room. The music therapist stands by the bed, talks directly to the patient and family, and gradually family members gather around the bed and begin interacting with each other and Dan. They respond to the statements by the music therapist, and a teenage son who was looking through a songbook states, "Dad always turned up the radio in the pickup when he heard this one." The music therapist suggests singing this song, and the family does so. The family begins telling stories about Dan, touching his arms and legs as they stand near the bed while others hold his hands. The music therapist encourages family to speak directly to Dan, as the sense of hearing is the last sense remaining before end-of-life. Dan may not be able to respond, but he possibly can hear what his family is saying to him. Family expresses a variety of emotions, sings, and talks to Dan, who responds by a slight raising of his eyebrows and barely noticeable nods of his head.*

This situation is typical in hospice and palliative care when a patient is nearing end-of-life. Other patients in hospice care are active, fully functional, and continue to work, while most others fit somewhere between on the continuum of nonresponsive to fully functional. In the scenario above, the music therapist provided a focus for the family members to interact with each other, to express emotions, and to express themselves directly to their loved one. The variety of emotions expressed was directly related to the music and the skills of the therapist who helped normalize the environment. Family members were able to laugh and shed tears when telling stories. They were able to review life and help put things in perspective while working together in their grief. The music was a way for the family to feel as if they could do something for Dan; they at least had control within this one area. The music therapist helped create positive memories in a situation where the family felt helpless and distressed.

Music therapy can assist with providing the focal point for the family to work together, everyone at the same time, for the best quality of life for the patient. Family can be encouraged to sing or play tone chimes or other instruments, working together as a cohesive unit for the benefit of the patient and everyone involved.

Music therapy also provides psychosocial support through music assisted relaxation, through music-based discussions for expression of concerns and fears, and by offering a focus for living in the present moment while enjoying the simple pleasures in life. Patients often find that a familiar meaningful piece of music may offer a calming and soothing presence. Songwriting, lyric discussions, singing, and listening to music can all assist patients and families with identification and expression of concerns or fears. Live music facilitated by the therapist brings the focus on the here and now, enjoying the moment with loved ones instead of worrying about the uncertainty of the future.

Provide emotional support. The music therapist has the means to reach patients on an emotional level (see Chapter 3). Instead of talking on an intellectual level, patients may be able to express themselves on the feeling level with assistance from the musical stimulus. Salmon (1993) maintains that most people have experienced "being profoundly moved upon hearing a piece of music" (p. 49).

Low-functioning patients may be able to cry, smile, or respond motorically, or express other emotions that would be more difficult or impossible to do with only discussion involved. Patients can be reassured that is it socially acceptable to release feelings to music, which may help normalize the situation. Often patients will express a certain emotion in reaction to music and not realize that this is what they were feeling. The music therapist can help identify and encourage expression of this feeling through music listening, songwriting, singing, instrument playing, lyric discussion, song choice, relaxation and music imagery, or making a recording as a lasting gift.

Music may be used to help someone visualize different ways of thinking or become self-aware, either through lyrics, music-based counseling, or self-growth within musical experiences. Patients may not need or desire extensive reframing or interpretation by the music therapist. Offering support and comfort through music-assisted supportive counseling and active listening when applicable may be most effective in providing the best quality of life for each individual.

As a stimulus for reminiscence/life review. Looking back at one's life and putting things in perspective helps a person discover a sense of meaning (Salmon, 1993). Music can effectively stimulate the long-term memory of patients, making it possible to recall in great detail long-forgotten past events and emotions (Bonny, 2001). This is particularly effective with patients who have short-term memory problems but retain all or part of their long-term memory. Families are able to contribute to a life review, enjoy, express emotions, and reconnect with each other through shared memories and experiences. Varied musical interventions can help the life review process become more vivid, detailed, and effective (O'Callaghan, 1996).

A musical life review can contain a mixture of emotions with patients and families expressing sadness, joy, hope, meaning, and release through the verbal process as well as the music itself. The musical presence in a life review tends to bring the emotional content to the surface, possibly making it more meaningful to those involved. For patients who enjoy singing, the release of tension, emotion, and creativity can be cathartic.

Provide spiritual support and comfort. The music therapist may work with the patient and family to select religious music according to specific faith traditions or with other music that is deemed spiritual by the patient or family. Patients may be physically unable to attend their place of worship or enjoy nature in a direct manner, so the music therapist can create opportunities for spiritual expression. Although spiritual needs are considerable and diverse, music therapists often provide comfort, a stimulus for reminiscence/life review, and an outlet for creativity (Hilliard, 2005; Krout, 2003; Trauger-Querry & Haghighi, 1999).

Music may be most comforting for a patient at end-of-life, as the bond between music, emotions, and spirituality is strong (Walker, 1995). Music may bring a sense of familiarity, intimacy, connectedness, tenderness, and peace as it blocks or masks other undesirable noise in the environment. Listening to favorite music from the past may be soothing for the patient, caregivers, and families. One person may wish to listen to favorite hymns sung quietly at bedside, which also helps reduce agitation later in the day. Another person may choose to hear upbeat gospel music that lifts mood and provides structure for motor responses such as toe-tapping and clapping. Both interventions provide spiritual support and involve passive or active music listening.

Music also brings the added dimension of comfort for patients who are not affiliated with organized religious practices. Offering comfort through musical pieces that have specific function or ritual can be particularly meaningful for patients and families. Music provides a venue for worship (Bonny, 2001). It helps access the deeper inner nature of being, opens communication between people and the divine, and provides structure for comfort, peace, and release (Lipe, 2002). Music, prayer, and the beauty of nature are effective means to access the close connection to a Higher Power for some patients (Wein, 1987).

Opportunities for choice and decision making. Patients may have little independence or autonomy at end-of-life. Singing and song choice can add increased control and self-expression to music listening approaches. The familiarity of choosing and singing even parts of a song seems to help memories become more vivid. Music therapists can encourage autonomy with simple choice-making interventions. For example, the music therapist might ask, "Do you want to hear 'Home on the Range' or 'As Time Goes By'?" This begins the process of enhancing feelings of control in a small manner. Patients make a selection, and then the music therapist can continue to offer an additional choice, "Would you like a fast song or a slow song?" and so on. Some patients who feel helpless need to practice making small decisions, and as they become more comfortable, they can make choices more easily.

The music therapist might provide the patient with an age-appropriate instrument, adapted so it is easily played in a successful manner, to enhance self-esteem and feelings of accomplishment. Nonverbal patients can often express themselves through motoric responses on a drum or other instrument. Songwriting can be adapted to ensure that a positive experience occurs and that autonomy will be enhanced. For example, a patient may need to fill in only one word or part of a phrase in a song that the music therapist has created. Discussion of song lyrics can help people identify strategies to develop realistic ways of taking control of things that can be controlled. Conversely, serious discussions of what types of things are beyond control can help validate experience and motivate change. The music therapist can help patients adjust to limitations, gain a sense of control, and raise self-esteem through interventions including song choice, music listening, songwriting, playing instruments, singing, making a recording for others, or lyric discussion.

Lower functioning patients may also participate in song choice and singing approaches. Allowing extra time for response, providing a forced choice between two selections, or having a visual aid helps some patients with making choices. For example, a nonverbal patient may be able to make a selection by pointing or directing a gaze at a picture of a sun, choosing the song You Are My Sunshine. Lower functioning patients may frequently join singing when adaptations such as repetition, slower tempo, lower range, and close proximity are implemented. They may be able to mouth words to songs, hum, and sing parts or ends of phrases with pleasure and a feeling of achievement.

> *Emily was a 6-year-old female with an inoperable brain tumor, no longer active due to disease progression. Music therapy provided sessions for*

> *Emily with her younger sibling and other family members. Emily preferred quiet voices and low stimulation due to her diagnosis, but wanted to sing and have everyone around her. She gained great comfort by singing her favorite song for others and from having her family repeatedly sing her favorite song to her. In addition to comfort, providing group sessions and a recording of Emily singing also helped create positive memories for Emily's younger sister and family.*

Singing, music listening, and song choice are adaptable to all ages as well as levels of functioning. In addition to their illness, children in hospice care may have the stress of missing school, events, friends, and the normal situations of everyday life. Adolescents in particular may listen to music for peer acceptance, to tune out adults, and for an emotional outlet. Music helps provide a normalized environment for children, offers distraction and a means for emotional expression, as well as providing comfort.

Outlet for creativity. Using music and creativity as a means of living fully and finding inner peace brings comfort at end-of-life (Krout, 2001). People who have life-limiting or chronic conditions are still able to express themselves creatively through music-based interventions in a variety of ways (Hilliard, 2001; O'Callaghan, 1996). Patients can sing, play instruments, write songs or poems that can be set to music, make recordings for lasting gifts, and make creative suggestions while participating in music therapy sessions.

Support during bereavement. Music therapy services can also be tailored to effectively support individuals in their own grief process (Krout, 2000). Each person has his or her own time frame for healing because there is no typical way to grieve. **Bereavement** begins for families, friends, and others of significance after the death of a loved one (Krout, 2005). Some music therapists terminate services upon death, others provide music and support at funerals or memorial services, and others continue to see family members for grief support for a length of time. Music therapists may also provide music for memorial services offered through the hospice organization for all patients who have died within a certain time period. It is common practice to offer bereavement sessions through the hospice organization for specific groups based on age or type of loss. For example, there are music therapy sessions for children who are bereaved, and grief groups for teens, parents, or spouses who have lost a partner.

Chapter II   The Difference Between Music and Music Therapy

This section aims to clarify the difference between the use of ***music*** as an intervention for pain perception and the use of ***music therapy*** as an intervention for pain perception.

2.1 Music as an Intervention

The studies conducted by a non-music therapist use recorded music only. The researcher often chooses the music as opposed to the patient, there is no relationship between the researcher and the patient, and there is no assessment of the patients' needs outside of alleviating pain perception. In these studies, the findings are a reflection of the act of 'listening to music'. These studies provide useful information about the tool of music and the effects of music on pain perception. Zimmerman et al., (1989) found that music and relaxation significantly decreases the overall pain experience in chronic pain patients. One of music's most documented medical uses is as an audioanalgesic, which has been described as taking aspirin through the ears (Mites, 1997 p.137).

Neurological pathways have been identified supporting music's ability to produce an endorphin response, which is the body's natural response to pain (Baker, 1998). Analyzed data displays significantly elevated pain thresholds with the use of soothing music (Whipple and Glynn, 1992) and that self-selected music can influence chronic pain perception. Testing for the confounding variable of medications revealed that medications do not influence how an individual responds to the intervention of listening to music. These findings are supported by research and clinical experience using music to assist patients with pain management in a variety of medical settings (Presner 2001). A study by Knight and Rickard (cited in Aldridge, 2003,p.17) supports the *anxiolytic[1] affect of music using a healthy student population.

There are three possible ways that music may modify pain (Magill-Levreault, 1993).

1. **Affective**: Music may alter mood disturbances associated with long term and life threatening illnesses such as anxiety, depression, fear, anger, and sadness. Music can lift depressive symptoms, promote relaxation, and thus diminish tension and anxiety.

2. **Cognitive**: Associative qualities of music provide a means of distracting attention away from pain often creating images and carrying a person's thoughts away from the noxious stimuli. Music provides a mechanism to improve patients' sense of control.

3. **Sensory**: Sensory component of music may have effect on sensory component of pain through counter-stimulation of the afferent fibers.

The act of listening to music may reduce the perception of pain as a distraction as used in a study for procedural pain and anxiety in patients with cancer. This kind of a distraction can change the transmission of pain impulses through activating the limbic system and sensory region of the brain (Kwekkeboom, 2003).

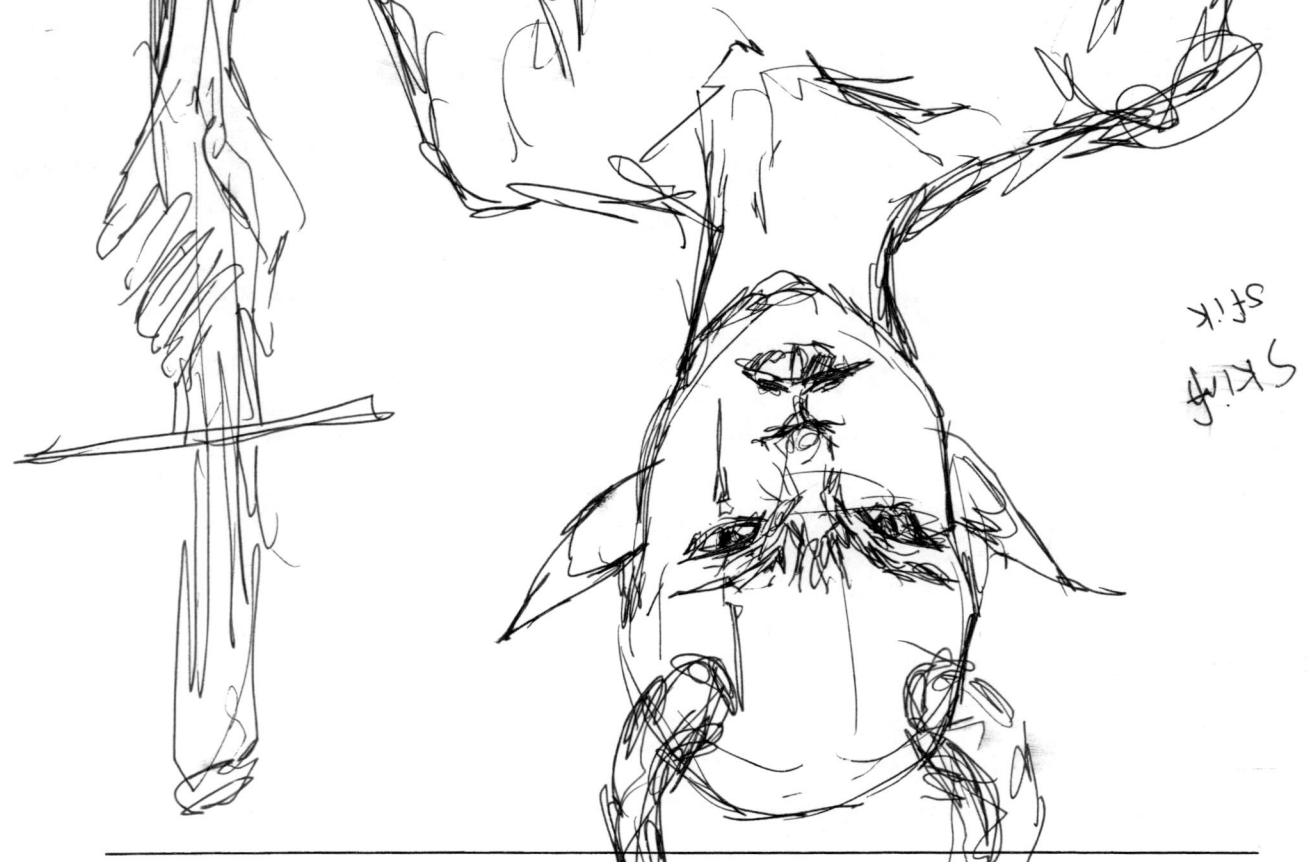

---

[1] *Anxiolytic: a drug prescribed for the treatment of symptoms of anxiety.

2.2 Music Therapy as an Intervention

Music therapists use the above noted attributes of music as their tools to reach client/patient goals. Research exploring music therapy and pain is undertaken by a qualified music therapist opposed to any other interested professional ie., nurse, doctor, physiotherapist.

As a music therapist, the music is administered in a variety of ways, often as a means of expression. There is a therapeutic relationship between the music therapist and the patient that develops during an assessment period and a treatment plan is created with the patient. The treatment plan covers goals and objectives inclusive of reducing the perception of pain. The tools implemented are fashioned to the individual and may involve a mix of pre-recorded music, live music, improvisation, composing lyrics, vocalization and/or instrumentation.

When patients are asked in Bullington's study (2003) to describe the path from seeking medical help to finding successful rehabilitation the main metaphor that arose was *order out of chaos*. Ordering chaos is regarded as a process moving from diagnosis through to a phase of heightened self-awareness. Related themes in Bullington et al's research concern the role of flexibility and creativity in the healing process and finding new meaning. These are integral elements to music therapy.

Music therapy studies are a reflection of a process involving music in a variety of forms. It is this *process* that patients identify a need for in Bullington's study. The music therapy process involves flexibility, creativity and meaning underlying the difference between using music in isolation and music therapy.

Surveys on manifestations of pain and suffering have led to the determination that the psychological component of pain is so important that modification of pain through psychological techniques is to be expected (Magill-Levreault, 1993). Music therapists perform comprehensive assessments that include reviews of social, cultural, medical history, current medical status and the ways in which emotions are affecting the pain.

Patients about to undergo surgical procedures are highly stressed; in such situations music therapy helps (Aldridge, 2003, p.17). A study at the Bristol Cancer Help Centre, UK, concludes that music therapy increases well-being and relaxation and decreases tension. The participants in this study had decreased levels of the hormone cortisol (Aldridge, 2003, p.19). High levels of cortisol are linked with stress and depression, which are directly linked to pain perception (Magill-Levreault, 1993). Music therapy can be an effective means to reducing anxiety as indicated in Kerr's study in 2001. This study compared typical reframing techniques as an intervention for reducing anxiety to reframing techniques with the addition of music by a music therapist. The results displayed music therapy as a more efficacious means to reducing anxiety (Kerr et al, 2001) Music therapy has also proven to be helpful as an intervention to those in intensive care (Aldridge, 2003, p.17).

2.3 Musical Interventions

For a better understanding of what kind of musical interventions may be used in music therapy the following is a list of interventions as provided by the CAMT (Canadian Association For Music Therapy). (http://musictherapy.ca, 2006)

1. *Singing*: Improves articulation, rhythm, and breath control. Improves oxygen saturation rates of individuals with COPD. Encourages reminiscence and discussions of the past, while reducing anxiety and fear.

2. *Playing Instruments*: Improves gross and fine motor coordination. Enhances cooperation, attention and provides opportunities for leadership-participant roles. Develops an increase in well-being and self-esteem.

3. *Rhythmic Based Activities*: Facilitates and improves upon range of motion, joint mobility/ agility/ strength, balance, coordination, gait consistency and relaxation. Rhythm and beat are important in "priming" the motor areas of the brain, in regulating autonomic processes such as breathing and heart rate, and maintaining motivation or activity level. Assists with receptive and expressive processing difficulties (ie.aphasia, tinnitus) to tolerate and successfully process sensory information.

4. *Improvising*: Offers a creative, nonverbal means of expressing thoughts and feelings. It is non-judgmental, easy to approach, and requires no previous musical training. Where words fail or emotions are too hard to express, music can fill the void. Where trust and interaction with others has been compromised due to abuse or neglect, improvisation provides a safe opportunity for restoration of meaningful interpersonal contact.

5. *Composing*: Facilitates the sharing of feelings, ideas and experiences. For people with a terminal illness, it is a vehicle for examining feelings about the meaning in life and death. It may also provide an opportunity for creating a legacy or a shared experience with a caregiver, child or loved one, prior to death. Lyric discussion and song writing can help individuals deal with painful memories, trauma, abuse, and express feelings and thoughts that are normally socially unacceptable, while fostering a sense of identification with a particular group or institution.

6 *Listening*: Helps to develop cognitive skills such as attention and memory. For those facing surgical procedures it allows the individual an opportunity to exert a sense of control over an often unpredictable environment. In situations where cognitive perceptions are compromised, such as in early to mid stage dementia, listening can provide a sense of the familiar, and increase orientation to reality. For those with psychological disorders such as Schizophrenia or Bipolar Disorder music listening can facilitate openness to discussion and provide motivation for increased levels of activity.

These music therapy interventions may be used individually in a session or jointly to reflect the needs of the patient. This research project has compiled research from all of the above listed interventions.

2.4 Benefits of Music Therapy As An Intervention For Pain Perception
(Finnerty, 2006)

The following chart was completed by the primary researcher to simplify how these interventions may effect one's perception of pain. All of the noted music therapy interventions result in a benefit of well being, relaxation, catharsis and/or distraction all of which have the potential to lessen pain perception. These benefits have been observed by the primary researcher as well as noted in the above section 2.1, referenced from the Canadian Association of Music Therapy.

MUSIC THERAPY INTERVENTION	BENEFIT	PAIN PERCEPTION
Playing Instruments	Increase well-being, feeling of being a part of something. Self expression	Well-being = Less pain
Rhythmic Based Activities	Regulate Breathing/Heart Rate. Relaxation	Relaxation = Less pain
Improvising	Meaningful Interpersonal Contact/Expression. Possible release of endorphins.	Catharsis = Less Pain Endorphins-body's natural pain killers
Listening	Stimulate Memories, Illicit Emotions, (stimulation of limbic system)	Overrides pain subjective experience = Less pain.
Composing	Sharing Feelings, Experiences, Possible release of endorphins (stimulation of limbic system)	Catharsis= Less pain Endorphins-body's natural pain killers

The aforementioned interventions also result in empowerment, another factor that influences pain perception. A patient can feel completely disempowered while staying at a hospital, finding it too painful to engage in daily activities such as washing, eating and sleeping. Daveson (2001) and Proctor (2002) have written articles in relation to music therapy and empowerment (as cited in Rolvsjord, 2004, p.100). It is suggested that empowerment is intrinsic to music therapy practice due to its participatory process and client ownership. Music making builds upon people's experiences of who they are and what they can do (Rolvsjord, 2004, p.100). Music therapy provides a vehicle of expression, which can be essential to pain relief, particularly for those patients who are inarticulate in the face of strong emotions (Aldridge, 2003). Participation in bedside sessions may begin with the music therapist singing in order to facilitate eye contact. The patient may then be encouraged to sing along, to move to the music (i.e., tap a finger, wiggle toes, wave a hand, nod the head) or to play some other instrument that requires little effort apart from active engagement through a structured, rhythmic response. This participation requires patients to focus on the musical productivity at that moment rather than the pain (http://healthpropress.com, 2002).

The benefits of music therapy in pain management are currently recognized at some health institutions such as; the Pain Service to the Neurology Department, New York, USA (Aldridge, 1996, p.75.).

Chapter III   Biological Parallels of Music Therapy and Pain

This section will highlight the biological parallels of music therapy and pain in order to gain an understanding of how music therapy may be influential in the perception of pain.

3.1 Subjective Experiences are Biological Events

As neuroscience maps the brain in more detail, the gap between the mind and body begins to narrow and to show itself as a false construct. The body begins to look smarter and more soulful (flesh as 'spirit thickened' as surgeon and author Richard Selzer has written) at the same time that the mind incarnates itself as a biochemical event (Jackson, 2002 p.14). Within the definition of pain there is an inherent divide between mind and body; the subjective and physical experience of pain. In recognizing how music therapy can effect pain perception, it is important to recognize that subjective experiences are physical. Subjective experiences are based on memories and emotions which are biological events controlled by the body's various communication systems.

3.2 The Body's Communication Systems

Research to date shows that the ability to perceive, feel, think, know and do is a result of neurons firing and resetting. Signals are generated and carried through networks of neurons. These signals pass through the nervous system; a communication system comprised of the central nervous system (CNS) and the peripheral nervous system (PNS). The CNS consists of the brain and the spinal cord. The somatic nervous system transmits information from the senses (sight, sound, smell, touch and taste) to the CNS, which then triggers neurotransmitters in the brain and muscles to move the skeleton (Koestler and Myers, 2002 p.13). The PNS is composed of the autonomic nervous system (ANS) and the somatic nervous system (SNS). The autonomic nervous system (ANS) causes the adrenal glands to release epinephrine (adrenalin) into the blood system. It is responsible for what is commonly known as 'fight or flight'.

3.2a Communication Systems
(Finnerty, 2006)

The following diagram was formulated by the primary researcher to display how the different communication systems interact.

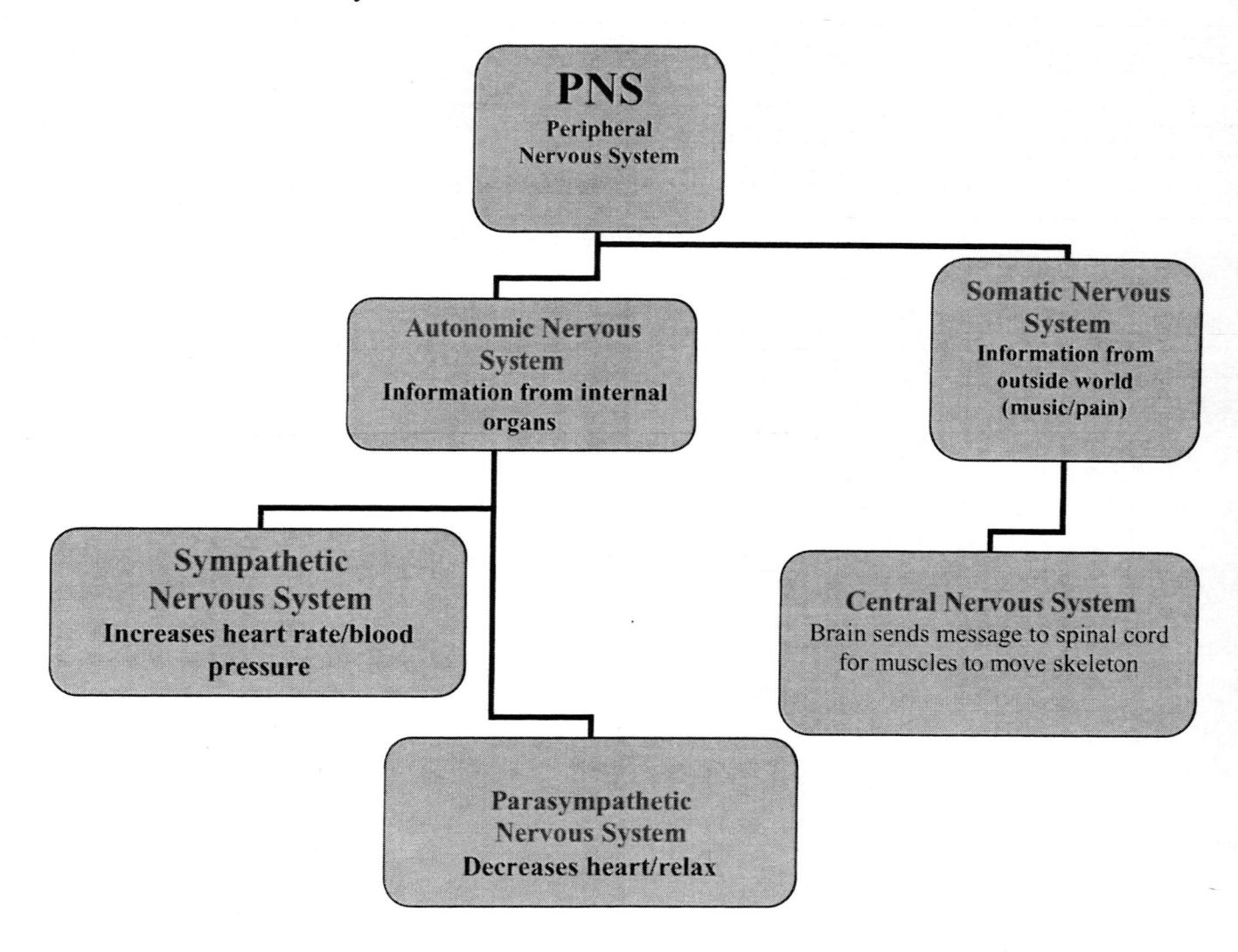

- **PNS** — Peripheral Nervous System
  - **Autonomic Nervous System** — Information from internal organs
    - **Sympathetic Nervous System** — Increases heart rate/blood pressure
    - **Parasympathetic Nervous System** — Decreases heart/relax
  - **Somatic Nervous System** — Information from outside world (music/pain)
    - **Central Nervous System** — Brain sends message to spinal cord for muscles to move skeleton

3.3 Neurotransmitters

Neurotransmitters are chemicals that are used to relay, amplify and modulate electrical signals between a neuron and another cell. The brain learns which nerves are transmitting information from particular areas of the body. Neurotransmitters that mediate pain (and emotions) are norepinephrine, dopamine, melatonin, epinephrine, L-dopa, serotonin, prolactin and enkephalines. A study by Kumar et al (1999) collected blood samples from 20 war veterans with Alzheimer's disease and observed an increase of melatonin, epinephrine and norepinephrine after 4 weeks of regular music therapy. Raised melatonin levels remained for a total of six weeks post sessions. Melatonin is largely known for preventing 'the blues'. The music therapy sessions in this study were inclusive of pre-recorded material only. Serotonin and prolactin levels were not affected in this study. Serotonin (5-HT) is a neurotransmitter that plays a role in regulating mood, sleep, vomiting, appetite and anxiety. Mood, sleep, appetite, and "the blues" are all elements that effect one's perception of pain and can be influenced by music.

Animal studies have shown that when serotonin is added directly to the CNS it accumulates in the periventicular areas and enhances the effectiveness of analgesia and decreases pain perception. When there is an abundance of serotonin being transmitted an individual typically feels a sense of well being and comfort. This is observed when MDMA (street drug Ecstasy) is implemented. MDMA is a drug that inhibits serotonin uptake, which causes a flood in the synapse (the space between neurons). It is also thought to raise levels of norepinephrine and dopamine. Blocking presynaptic reuptake of serotonin in the synaptic cleft raises the threshold for pain. When serotonin is administered as a pharmaceutical it is unable to cross the blood brain barrier preventing it from directly affecting the CNS, disenabling its effectiveness for pain relief unless taken with a synthetic precursor. This is the same for dopamine (DA). Dopamine is commonly associated with the *pleasure system* of the brain, providing feelings of enjoyment, contributing to reinforcement and motivation.

Listening to music can affect the production of dopamine (DA) in the brain by stimulating the production of serum calcium; this was discovered in a study by Sutoo & Akiyama (2004). Previous studies have indicated that calcium increases brain dopamine synthesis, which in turn affects blood pressure. When rats are exposed to music (in this case Mozart's K.205), their systolic blood pressure is reduced and calcium levels are 5-6% higher than non-music control rats (Sutoo & Akiyama, 2004).

As well as neurotransmitters, dopamine and serotonin are hormones. Hormones act as "messengers," and are carried by the bloodstream to different cells in the body, which interpret these messages and act on them.

3.4 Manufactured Pain killers

Manufactured pain killers are made to act on neurotransmitters cutting down the messages passed from cell to cell (Wells and Nown, 1998 p.13). Pain killers are referred to patients by doctors and pharmacists on an analgesic ladder. Over the counter drugs are rung 1, codeine is rung 2, and morphine is rung 3. Rung 2 and 3 are controlled substances. When an injury occurs, the damaged tissue releases the hormone prostaglandis (PG), which stimulates nerve endings to carry pain messages to the brain (Wells and Nown, 1998). Drugs such as acetylsalicylic acid (ASA or Aspirin) inhibit the production of PG resulting in the reduced sensation of pain. Patients who are elderly have a lower toleration for pain medications because aging causes physiological changes that alter the pharmacokinetics and pharmacodynamics of analgesics. This narrows their therapeutic index, and increases the risk of toxicity and drug-drug interactions (Davis and Srivastava 2003). Many manufactured pain killers are addictive, such as morphine, codeine like drugs (if used for more than 6 months) and tranquilizers (benzodiazepine, diazepam, lorazepam) intended to work as a muscle relaxant, addiction can occur after 6 weeks of use (Wells and Nown, 1998).

3.5 Nociception

Nociceptors are the nerve cells that transmit unpleasant impulses to the brain. Nociception is a neurological term that refers to the transmission mechanism of physiological pain. In response to a pain message from a part of the body, the brain can order the release of chemicals that reduce or inhibit pain sensations (Wells and Nown, 1998, p.7). These pain blocking chemicals produced by the body, are called endorphins and are the key to controlling pain. It has been theorized that music might alter pain through affective and cognitive effects that stimulate endorphin production and the other endogenous mechanisms for pain modulation (Wells and Nown, 1998).

3.6 The Gate Control Theory

Gate Control Theory (GCT) proposes that activity in the large afferent nerves may inhibit activity in the smaller fibres resulting in a reduced perception of pain. For example, rubbing a bumped knee seems to relieve pain by preventing its transmission to the brain. It is possible that music can affect the gating mechanism through the suppression of signals descending from the brain to the spinal cord.

One end of each cell has a receiver that picks up a pain message, and the other end has a chemical substance that passes on the pain message to stimulate the next cell. The brain depends on T-cells (central transmission cells) to send messages from the dorsal horns of the spinal cord. Wall and others suggest that the afferent barrage arriving in the spinal cord can be processed at one of three different speeds. The speed of this information processing depends on the context of other information coming in from the periphery and the integration of descending messages from the brain (Skevington, 1995 p.14). Messages are modified at the gates as a result of emotions and what is happening within the body at that given time. Information can be processed either rapidly or assimilated more slowly (minutes or hours rather than seconds) to result in sensitivity control.

Music therapy may be influential in affecting the speed of processing information about pain, modifying the message at the gate by affecting emotions and mental state.

3.7 The Limbic System

The limbic system operates by influencing the endocrine system and the autonomic nervous system (ANS). It is a complex set of structures that lie on both sides and underneath the thalamus and under the cerebrum. These structures are largely responsible for emotional life. They are intensively interconnected and none is solely responsible for any specific emotional state (Rocha do Amaral and Martins de Oliveria). It is believed that connections between centres in the brainstem such as the reticular formation and the limbic system account for strong unpleasant emotions reported by those experiencing pain. The prefrontal and limbic brain structures have been noted to be involved during music therapy sessions (Esch et al, 2004). Particularly the left anterior regions of the brain, which are involved in reward and motivation. Listening to music engages the limbic and paralimbic systems, the same systems involved in pain perception. This is observed in studies using neuroimaging and PET scans (positron emission tomography) while an individual passively listens to music.

3.7a Limbic System

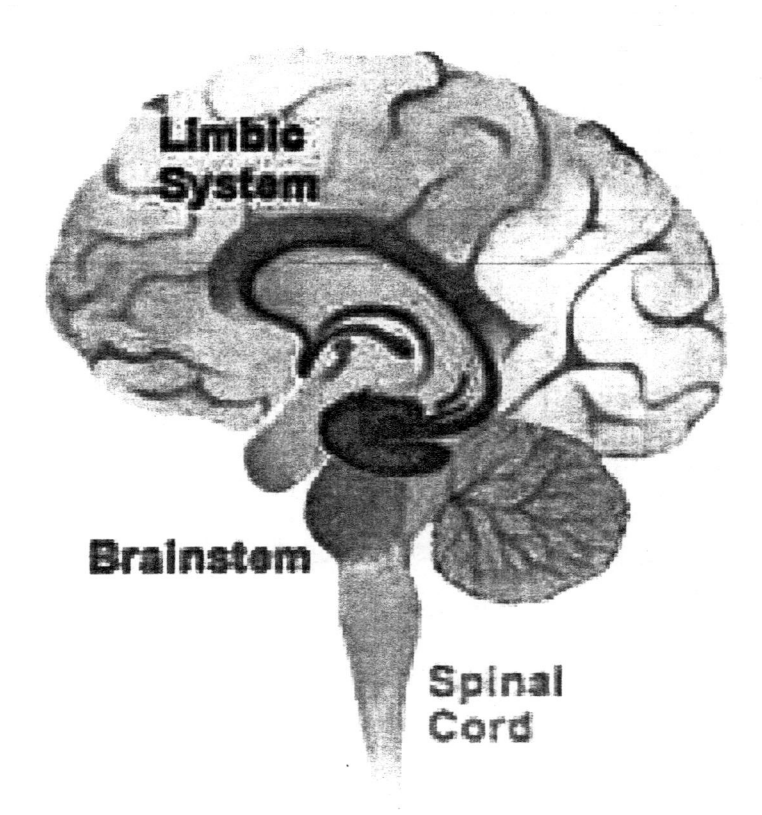

http://en.wikipedia.org/wiki/Limbic_system 2006

3.7b Limbic System-Structures and Relevance to Pain and Music Therapy
(Finnerty, 2006)

There is some controversy over which structures are to be named as part of the limbic system. The structures listed below are the generally accepted structures of the limbic system. The role of each limbic structure is cited from Martins de Oliveria and Rocha do Amaral. For the purpose of this research project, the primary researcher has categorized the relevance of each structure with the relevant research about pain and about music therapy as an intervention to pain.

*In the categories to follow, music therapy is referred to as MT.

Hypothalamus
Mainly concerned with homeostasis regulating hunger, thirst, response to pain, pleasure, sexual satisfaction, anger and aggressive behaviour. It also regulates the parasympathetic and sympathetic nervous system, which in turn regulate the pulse, blood pressure, breathing and arousal in response to emotional circumstances.
Relevance to Pain
The parasympathetic and sympathetic nervous system are responsible for the "fight or flight" response. Pulse rate, blood pressure and breathing increase as a result of anxiety provoking stimuli such as variable surrounding the experience of pain.
Relevance to MT intervention
Music is able to influence the parasympathetic and sympathetic nervous system by decreasing pulse rate, blood pressure and breathing bringing the individual into a relaxed state.
Mammilary body & Hippocampus
Important for the formation of memory. Converts short term memory to long term memory. If it is damaged, new memories cannot be stored.
Relevance to Pain
Memories provide individuals with their subjective experience of pain.
Relevance to MT intervention
Music is able to evoke memories. Evoking pleasurable memories and a pleasurable experience may provide a new subjective experience in relation to pain and override the memories that enhance pain perception.
Nucleus Accumbens
Involved in reward, pleasure and addiction.
Relevance to Pain
Pain inhibits reward/pleasure systems.
Relevance to MT intervention
Music evokes feelings of pleasure that may act as a distraction from the feelings associated with pain.

30

Orbitofrontal Cortex:
Required for decision making.
Relevance to Pain
Patients' ability to make appropriate decisions for improved quality of life may be impaired due to loss of motivation as a result of pain.
Relevance to MT intervention
Music therapy provides opportunities for the patient to make decisions that will ultimately reduce pain perception through building self-confidence and motivation.
Ventral Tegmental Area
Consists of dopamine pathways that are believed to be responsible for pleasure.
Relevance to Pain
Pain killers are manufactured to act upon dopamine pathways to decrease pain perception
Relevance to MT intervention
Studies have suggested that music therapy increases levels of dopamine (Sutoo and Akiyama 2004). This contributes to the individual's experience of pleasure and motivation, which is essential for an individual to maintain a pain management regime.
Cingualted Gyrus
Participates in the emotional reaction to pain and in the regulation of aggressive behaviour. In extreme cases of chronic pain, a patient may undergo an anterior cingulectomy. This procedure disconnects the anterior cingulated gyrus allowing the patient to continue to feel the sensation of pain without the accompanying emotion. This procedure highlights the importance of the emotional aspects of pain to the quality of life.
Relevance to Pain
The emotional aspects of pain provoked by a change of lifestyle and daily activities may become overwhelming contributing to diagnosed depression.
Relevance to MT intervention
Music therapy is able to influence emotional states and provide an opportunity to process the emotions evoked by pain.

3.8 The Limbic System and Emotions

Emotions are an integral part of the experience of pain. Patients are likely to have an emotional response to the noisy, bustling environment compounded with a newfound lack of independence as expert caregivers, machines, and medications impose their directives and controls (Lusk and Lash, 2005). Each of these emotional factors has an impact on the perception of pain. Music therapy provides patients with a space to express their emotions decreasing the impact of the negative emotional factors.

Music is able to elicit emotional responses and fulfill expectations without the use of rational information processing (Esch at al, 2004). This makes music a particularly powerful tool in eliciting emotions by providing a vessel to enjoying musical moments despite rational information processing that is overwhelmed with the negative emotions associated with pain.

Emotions (affective pain) share the same pathway as the sensation of pain. These pathways (also shared by music therapy) are inclusive of related memory processing, accompanied by endocrinologic and autonomic functions. Through employing a PET scan (positron emission topography), Blood and Zatorre (2001) found that consonant music stimulates parts of the brain associated with pleasure and patients accordingly report positive emotions whilst listening to this type of music. Dissonant music stimulates parts of the limbic system associated with unpleasant emotions. Pleasant and unpleasant emotions are integral to the perception of pain.

Underlying neural mechanisms can be identified for intensely pleasant emotional responses to music. A PET study implemented by Blood and Zatorre (2001) with individuals who chose emotion- provoking music indicates that cerebral blood changes can be measured. These cerebral blood changes mark a link between music and biologically relevant survival stimuli. There is a common recruitment of activity in the brain known to be involved in the pleasure and reward circuit. Systematic research concerning music and emotions is minimal but there is an intersection with the literature on neuroscience of affective processing which is directly related to the experience of pain (Zatorre, 2003). This intersection highlights the impact that music and emotions (key elements to music therapy) have on pain. Through the use of self-administered scales for stress level and testing systolic blood pressure to measure heart rate, music therapy research has documented the use of music to reduce stress and heart rate. Blood and Zatorre (2001) confirm that the brain is active in the same areas during music therapy as when responding to pain beyond nociception.

168